Wolfgang Günthner

Enhancing Cognitive Assistance Systems with Inertial Measurement Units

Studies in Computational Intelligence, Volume 105

Editor-in-chief
Prof. Janusz Kacprzyk
Systems Research Institute
Polish Academy of Sciences
ul. Newelska 6
01-447 Warsaw
Poland
E-mail: kacprzyk@ibspan.waw.pl

Further volumes of this series can be found on our homepage: springer.com

Vol. 81. K. Sridharan
Robotic Exploration and Landmark Determination, 2008
ISBN 978-3-540-75393-3

Vol. 82. Ajith Abraham, Crina Grosan and Witold Pedrycz (Eds.)
Engineering Evolutionary Intelligent Systems, 2008
ISBN 978-3-540-75395-7

Vol. 83. Bhanu Prasad and S.R.M. Prasanna (Eds.)
Speech, Audio, Image and Biomedical Signal Processing using Neural Networks, 2008
ISBN 978-3-540-75397-1

Vol. 84. Marek R. Ogiela and Ryszard Tadeusiewicz
Modern Computational Intelligence Methods for the Interpretation of Medical Images, 2008
ISBN 978-3-540-75399-5

Vol. 85. Arpad Kelemen, Ajith Abraham and Yulan Liang (Eds.)
Computational Intelligence in Medical Informatics, 2008
ISBN 978-3-540-75766-5

Vol. 86. Zbigniew Les and Mogdalena Les
Shape Understanding Systems, 2008
ISBN 978-3-540-75768-9

Vol. 87. Yuri Avramenko and Andrzej Kraslawski
Case Based Design, 2008
ISBN 978-3-540-75705-4

Vol. 88. Tina Yu, David Davis, Cem Baydar and Rajkumar Roy (Eds.)
Evolutionary Computation in Practice, 2008
ISBN 978-3-540-75770-2

Vol. 89. Ito Takayuki, Hattori Hiromitsu, Zhang Minjie and Matsuo Tokuro (Eds.)
Rational, Robust, Secure, 2008
ISBN 978-3-540-76281-2

Vol. 90. Simone Marinai and Hiromichi Fujisawa (Eds.)
Machine Learning in Document Analysis and Recognition, 2008
ISBN 978-3-540-76279-9

Vol. 91. Horst Bunke, Kandel Abraham and Last Mark (Eds.)
Applied Pattern Recognition, 2008
ISBN 978-3-540-76830-2

Vol. 92. Ang Yang, Yin Shan and Lam Thu Bui (Eds.)
Success in Evolutionary Computation, 2008
ISBN 978-3-540-76285-0

Vol. 93. Manolis Wallace, Marios Angelides and Phivos Mylonas (Eds.)
Advances in Semantic Media Adaptation and Personalization, 2008
ISBN 978-3-540-76359-8

Vol. 94. Arpad Kelemen, Ajith Abraham and Yuehui Chen (Eds.)
Computational Intelligence in Bioinformatics, 2008
ISBN 978-3-540-76802-9

Vol. 95. Radu Dogaru
Systematic Design for Emergence in Cellular Nonlinear Networks, 2008
ISBN 978-3-540-76800-5

Vol. 96. Aboul-Ella Hassanien, Ajith Abraham and Janusz Kacprzyk (Eds.)
Computational Intelligence in Multimedia Processing: Recent Advances, 2008
ISBN 978-3-540-76826-5

Vol. 97. Gloria Phillips-Wren, Nikhil Ichalkaranje and Lakhmi C. Jain (Eds.)
Intelligent Decision Making: An AI-Based Approach, 2008
ISBN 978-3-540-76829-9

Vol. 98. Ashish Ghosh, Satchidananda Dehuri and Susmita Ghosh (Eds.)
Multi-Objective Evolutionary Algorithms for Knowledge Discovery from Databases, 2008
ISBN 978-3-540-77466-2

Vol. 99. George Meghabghab and Abraham Kandel
Search Engines, Link Analysis, and User's Web Behavior, 2008
ISBN 978-3-540-77468-6

Vol. 100. Anthony Brabazon and Michael O'Neill (Eds.)
Natural Computing in Computational Finance, 2008
ISBN 978-3-540-77476-1

Vol. 101. Michael Granitzer, Mathias Lux and Marc Spaniol (Eds.)
Multimedia Semantics - The Role of Metadata, 2008
ISBN 978-3-540-77472-3

Vol. 102. Carlos Cotta, Simeon Reich, Robert Schaefer and Antoni Ligeza (Eds.)
Knowledge-Driven Computing, 2008
ISBN 978-3-540-77474-7

Vol. 103. Devendra K. Chaturvedi
Soft Computing Techniques and its Applications in Electrical Engineering, 2008
ISBN 978-3-540-77480-8

Vol. 104. Maria Virvou and Lakhmi C. Jain (Eds.)
Intelligent Interactive Systems in Knowledge-Based Environment, 2008
ISBN 978-3-540-77470-9

Vol. 105. Wolfgang Günthner
Enhancing Cognitive Assistance Systems with Inertial Measurement Units, 2008
ISBN 978-3-540-76996-5

Wolfgang Günthner

Enhancing Cognitive Assistance Systems with Inertial Measurement Units

With 94 Figures and 5 Tables

 Springer

Wolfgang Günthner
University of California, Berkeley
2299 Piedmont Ave.
Berkeley, CA 94720
USA
guenth@me.berkeley.edu

ISBN 978-3-642-09572-6   e-ISBN 978-3-540-76997-2

Studies in Computational Intelligence ISSN 1860-949X

Cover design: Deblik, Berlin, Germany

Printed on acid-free paper

9 8 7 6 5 4 3 2 1

springer.com

Acknowledgements

I wish to gratefully thank Professor Heinz Ulbrich for his excellent supervision during my time at the Institute of Applied Mechanics. His continuous support of my ideas made it a pleasant and fruitful period. Many thanks are due to Professor Masayoshi Tomizuka for supporting my research at UC Berkeley. Professor Friedrich Pfeiffer has made an important contribution with his helpful discussions.

I would like to pass on my sincere gratitude to Professor Bernd Heißing and Professor Frank Schiller for serving on my committee as thesis reader and chairman respectively.

A special mention is due to Dr. Stefan Glasauer for engaging in many valuable debates on the neuroscientifc aspects of my work, and to my former office mate Philipp Wagner for his efficient cooperation. The work with all partners made FORBIAS a successful project. I would particularly like to acknowledge my industrial partners *BMW*, *Audi* and *Eurocopter* for their excellent and professional cooperation.

I am grateful to my lab-mate Lucas Ginzinger for his in-depth scientific proof-reading.

It is a pleasure to thank Dr. Katherine Richert, Luke Harley, Zoey Rothenberg, Fatma Hassan and Anneke Pot for their useful comments on the manuscript.

Many students decided to undertake research for their Bachelor's or Master's theses under my supervision and contributed to this work. All did a great job. Special mentions go to Chen Zhao and Christian Alt, who have been working with me for more than two years. I would like to thank Georg Mayr, Wilhelm Miller, Walter Wöß, Simon Gerer and Philip Schneider for their high-precision manufacturing of electronic and mechanical components.

This research has been funded by Bayerische Forschungsstiftung and a Heinz Dürr Scholarship. I am fortunate to have received support from the German National Academic Foundation *(Studienstiftung)* throughout my graduate studies. Marius Spiecker gen. Döhmann has also provided valuable advice.

Finally, I would like to thank my family who offered gentle counsel and unconditional support at each fork in the road. This book is the result of their fostering my curiosity and interest in science and technology from early on.

Berkeley, CA
November 2007

Wolfgang Günthner

Contents

1 Introduction

1.1 Motivation

Throughout the last decade, major car manufacturers have introduced a great variety of different assistance systems to make driving safer and more comfortable. Mechatronic systems like dynamic stability control, situation dependant suspension adjustment or the recently introduced night vision cameras are just a few examples among a range of helpful developments.

The current focus of research is on cognitive assistance systems for high-level perception of the environment. Rather than present mere images to the driver, these future systems extract all relevant information and provide a reduced set that alerts to possible dangers or, in a further stage, directly takes over control of the vehicle to prevent accidents. These *Steps Towards the Seeing Car* were presented as a keynote speech at the *Advanced Microsystems for Automotive Applications 2007* conference by VOLKSWAGEN [87]. In addition to radar, ultrasonic and laser sensors, concepts include video cameras for the perception of the medium and far range.

Video will play an especially important role in future driver assistance, since the traffic environment is designed for human needs, and thus relies predominantly on visual cues. For the visual assessment of the environment, two aspects are crucial and will be covered in this book. First, the vehicle has to know its own motion in space in order to reliably differentiate visual motion caused by other objects, such as moving pedestrians, from that of the static environment generated by the car's own movement. Second, the visual sensor needs to cover a wide range with high resolution and precision.

The first topic is best addressed using inertial measurement systems. While multi-antenna differential GPS can provide a good estimate of the vehicle's position, and to a certain extent its orientation, its low update rate makes it unsuitable for high frequency motion, especially fast vehicle rotation. To cover this motion, the tendency is to progress from single-axis sensors for yaw and roll (currently used for stability control), to a complete 6 degrees of freedom inertial measurement unit (IMU), which not only captures rotations in 3 DOF, but also transient translations. Sensor measurements can then be fused with GPS measurements to allow for better estimates of the orientation and position of the vehicle. The main challenge in the development of a competitive IMU is to provide a reliable low-cost device with a

W. Günthner: *Enhancing Cognitive Assistance Systems with Inertial Measurement Units*, Studies in Computational Intelligence (SCI) **105**, 1–5 (2008)
www.springerlink.com

suitable size for installation within the vehicle at any place. FIAT Research recently published its effort to develop an integrated IMU for orientation estimation [27].

The visual sensor topic can be best addressed with a hybrid system consisting of a wide-angle camera for the general overview, and a second pivotable camera with a telephoto lens for high resolution. However, the effectiveness of the second camera is challenged by the fact that images taken with high focal length, and thus a small aperture angle, are extremely sensitive to motion and vibration. This problem not only exists in technical systems, but also within visual systems in animals and humans. To solve this problem, nature developed gaze stabilization and the so-called vestibular-ocular reflex: in vertebrates, from fish to humans, inertial sensors in the inner ear measure head movement and control eye muscles in order to stabilize the image on the retina. Analogously, the same inertial measurement unit used previously to determine the car's movement can be the sensor for active camera stabilization. It is therefore part of this work to use knowledge from neuroscience for the technical application of camera stabilization.

1.2 Position of the Project within the Research Cooperation

The work covered in this book is part of a project embedded in the research cooperation for *Biomimetic Assistance Systems (FORBIAS)*, funded by *Bavarian Research Foundation (BFS)*. The goal of FORBIAS is to develop technical assistance systems and implement solutions that the human system already incorporates in nature for similar problems. A close cooperation with neuroscientists from the Center for Sensorimotor Research at the Department of Neurology of LMU Munich enabled its medical expertise to be brought into the project. In total, about a dozen researchers, most of them Ph.D. candidates, worked on the 4 million euro projects. In order to guarantee relevance for industry and the direct transfer of knowledge, half of the project funds were provided by major companies like AUDI, BMW, EUROCOPTER and SIEMENS, as well as by innovative small companies like JUMA PCB GMBH.

Within the cooporative, assistance applications headed in two different directions. Researchers at the LMU Munich University Hospital focused on a pivotable head-mounted camera, in which the wearer drives the camera simply by looking at a point, with the camera following. This technology is helpful for documenting surgeries, for example. Since the head camera system only uses inertial sensors but is otherwise developed independently, it only affects the sensor specification section of this book.

The second part, pertaining specifically to this work, addresses assistance systems in automotive applications, extending to aviation. At the outset, the inertially stabilized camera platform was intended only for use in driver assistance systems.

However, during the work on the project, the stabilized camera sparked the interest of EUROCOPTER, prompting an additional platform to be customized for pilot assistance in helicopters.

Fig. 1.1 summarizes the interaction of this project, originally only meant to develop an inertial measurement unit, with the two other relevant components *Camera Motion Device (CMD)* and *Scene interpretation*, directed towards industrial applications. The CMD was developed at the Institute of Applied Mechanics by WAGNER [135]; for the Scene interpretation algorithm, NEUMAIER from the Institute for Real-Time Computer Systems was responsible. The part covered in this book and shaded in Fig. 1.1 includes the development of the inertial measurement unit, adaptive control algorithms for inertial stabilization and target pursuit, as well as both application scenarios for driver and pilot assistance.

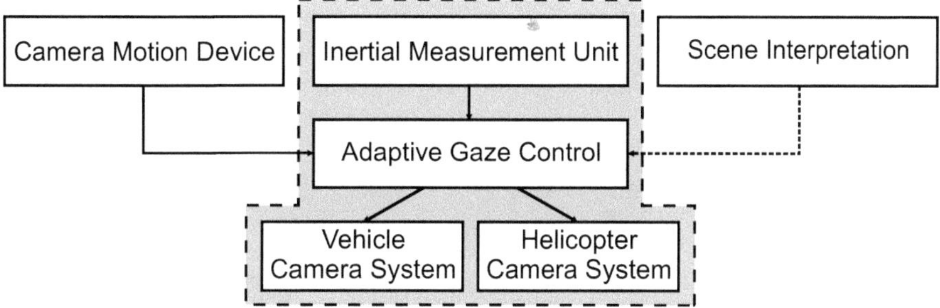

Fig. 1.1: Relevant excerpts from the FORBIAS project structure with topics covered in this book highlighted. Head camera system omitted for clarity.

1.3 Goals and Organization of the Book

The primary goal of this book is the specification, design and testing of an inertially stabilized camera platform for assistance systems with the focus on adaptive inertial measurement. This can be divided into the following sub-goals, which also served as internal milestones for the project:

- **Development of a highly miniaturized inertial measurement unit**
  Commercially available units are far too large for the intended application. For the camera platform to be in the immediate vicinity of the rearview mirror, the required IMU needs to have a design envelope small enough to even fit inside the mirror.

- **Development of adaptive control algorithms for gaze stabilization**
  To compensate for inevitable inaccuracies and manufacturing tolerances, an adaptive algorithm is required to avoid costly high-precision system identification procedures.

- **Industrial application**
  In the automotive application, the inertially stabilized camera has to track
  Earth-fixed, as well as moving, targets like other cars, while compensating for
  the test vehicle's self-motion. A stable high-resolution image of the point of
  interest is the ultimate goal. In the helicopter application, the aim is to deliver
  to the pilot a stable image of the landing place during final approach.

In addition to these camera-related goals, the inertial measurement unit shall be
used for orientation determination:

- **Development of multi-sensor fusion algorithms**
  The determination of motion and orientation in space is important for many
  applications. For future driver-assistance applications based on high-level terrain maps, suitable algorithms are needed to compute the orientation from
  information drawn from the developed inertial measurement unit. The focus
  in that section lies on accuracy in connection with low-cost sensors.

The structure of this book mainly follows the above mentioned goals and is laid out
in Fig. 1.2. In *Chapter* 2 the foundation for the successive part is established, the
state-of-the-art clearly outlined. This includes technical aspects of sensor technology and camera stabilization, as well as a neuroscientific approach to the human
equivalents.

In *Chapter* 3 the inertial measurement unit itself is developed. Focus is placed on
the combination of highest possible miniaturization in connection with competitive
system performance. Experimental sensor characterization completes this section.

Multi-sensor fusion algorithms are addressed in *Chapter* 4. Several different algorithms with increasing computational complexity are developed and compared to the
output of the inverse kinematics of a parallel robot. Numerical optimization further
increased the accuracy.

Chapter 5 focuses on the basic principles of camera stabilization. The first part, with
algorithms later used for industrial applications, is directed towards a multi-layer
system where the intended line of sight (LOS) is determined by a superordinate system, such as a wide-angle camera based environment model. The second section then
uses a unifocal system with only one stabilized camera, with the LOS determined directly from that camera image. Both approaches are evaluated on a Steward-Gough
platform.

In *Chapter* 6 the applications in cooperation with the project's industrial partners are presented. First of all the setup for cognitive driver assistance systems is
addressed. The core of this section then relates to test drives conducted under every-
day conditions on German freeways. The quality, as well as the benefits of active
stabilization of a camera with telephoto lens for the perception of the environment,
is discussed. Secondly, in a slightly modified application, a camera platform for pilot assistance during instrument-aided approach scenarios is developed. Test flights

undertaken in cooperation with EUROCOPTER show the operability of the system under technically challenging conditions. The section concludes with the evaluation of experimental results and recommendations for future applications.

Finally, Chapter 7 summarizes the research and results presented in this book.

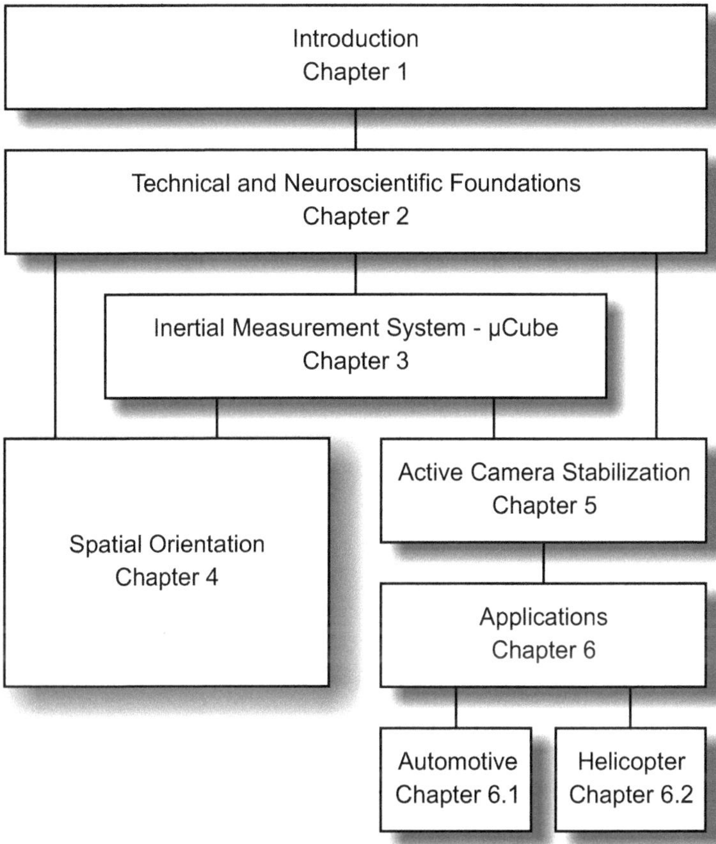

Fig. 1.2: Chapter structure of this book.

2 Scientific State-of-the-Art

2.1 The Human Vestibular System

For the derivation of the head and body's spatial orientation and motion, humans and other vertebrates possess the vestibular organs, which is the biological equivalent to a strap-down inertial measurement unit. Located in the two inner ears, these organs have overall dimensions of about 5mm. They are divided into two main parts for inertial sensing, Fig. 2.1.

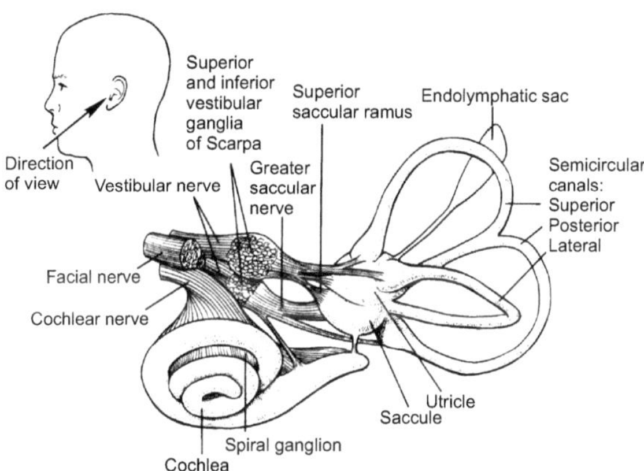

Fig. 2.1: Schematic drawing of the human vestibular organ [83].

The otholitic system is responsible for sensing linear accelerations. Utricle and saccule each are composed of three layers: the otoconial layer consisting of calcium carbonate crystals, the mesh layer, and the viscoelastic gel layer. Due to the density difference with respect to the surrounding fluid, the otoconial layer will move in response to acceleration in the horizontal and vertical plane respectively [91]. Displacement is detected by the deflection of hair cells raising from the macula into the gel layer, Fig. 2.2.

The resulting signal is transmitted via the vestibular nerve to the brain. The dynamic behavior of the otolithic system can be described with a low-pass characteristic. Young and Meiry developed a model using the transfer behavior

W. Günthner: *Enhancing Cognitive Assistance Systems with Inertial Measurement Units*, Studies in Computational Intelligence (SCI) **105**, 7–23 (2008)
www.springerlink.com   © Springer-Verlag Berlin Heidelberg 2008

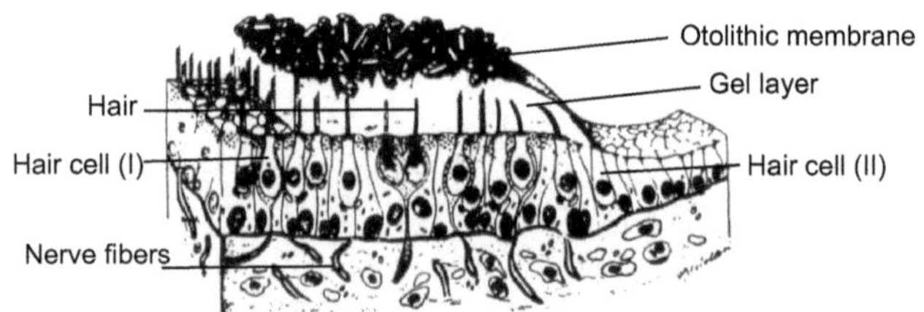

Fig. 2.2: Schematic drawing of a macula [91].

$$G_{o,Young}(s) = \frac{a_e}{a_o} = \frac{F(\tau_1 s + 1)}{(\tau_2 s + 1)(\tau_3 s + 1)} \tag{2.1}$$

between the externally applied acceleration a_e and the otolith output a_o with the Laplace operator s, the parameter $F = 0.4$ and the time constants $\tau_1 = 13.2$, $\tau_2 = 5.33$ and $\tau_3 = 0.66$.

In a later experimental study, Telban [128] linked the afferent firing rate (AFR) of the corresponding nerves to output

$$G_{o,Telban}(s) = \frac{a_e}{a_o} = \frac{AFR}{a_o} = \frac{F(\tau_1 s + 1)}{(\tau_2 s + 1)(\tau_3 s + 1)} \tag{2.2}$$

in the same structure, however concluding with a different parameter and time constants $F = 33.3$, $\tau_1 = 10$, $\tau_2 = 5$ and $\tau_3 = 0.016$. This does not change the form of the transfer function but only the position of the characteristic frequencies.

Angular rates are sensed by the semicircular canals of the vestibular organ. The three ducts (Fig. 2.3) with a diameter of about $0.8mm$ and a length varying from $12mm$ for the horizontal canal to $22mm$ for the posterior semicircular canal [62] are arranged in approximately orthogonal planes.

When exposed to angular acceleration, the fluid inside the canals (endolymph) is set into motion and bends the cupula. The attached hair cells of the macula again generate the sensory signal. Due to the small diameter of the canals, the viscous properties of the endolymph become predominant and deflection of the cupula is in phase with angular velocity for typical head motion. The canals perform an integration of the accelerations and thus the output to the brain is directly proportional to angular rates. Using information from the three ducts, rotation in space can be reconstructed.

Steinhausen (1931) and Egmond (1949) [35] made an analogy in between the semicircular canals and a highly damped torsional pendulum in the dynamic modeling. Both systems are governed by the second-order differential equation

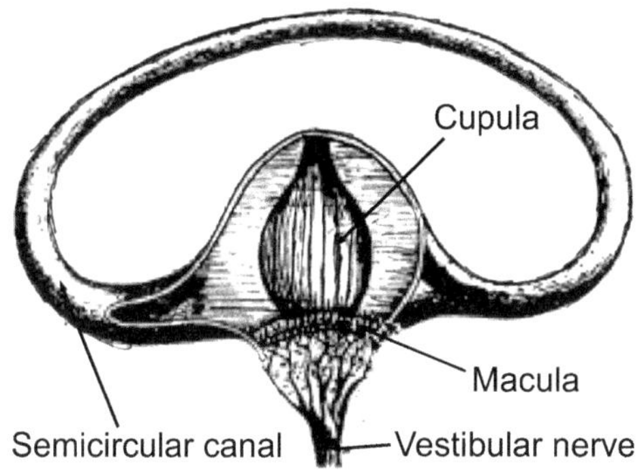

Fig. 2.3: Schematic drawing of one semicircular canal.

$$\Theta\frac{d^2\xi(t)}{dt^2} + \Pi\frac{d\xi(t)}{dt} + \Delta\xi(t) = \Theta\alpha(t) = \Theta\frac{d\omega(t)}{dt} \tag{2.3}$$

with

Θ moment of inertia of the endolymph
Π moment of friction at unit angular velocity
Δ directional momentum at unit angle caused by the cupula
ξ angular deviation of the endolymph in relation to the skull
α angular acceleration of the skull
ω angular velocity of the skull

In order to derive the transfer function of the semicircular canals, Eq. 2.3 is Laplace transformed

$$\Theta s^2\Xi(s) + \Pi s\Xi(s) + \Delta\Xi(s) = \Theta s\Omega(s) \tag{2.4}$$

$$\Xi(s) = \frac{\Theta s}{\Theta s^2 + \Pi s + \Delta}\Omega(s) \tag{2.5}$$

with the transforms of the variables

$$\Xi(s) = \mathscr{L}(\xi), \quad \Omega(s) = \mathscr{L}(\omega) \tag{2.6}$$

and the boundary conditions

$$\frac{d^2\xi(0)}{dt^2} = \frac{d\xi(0)}{dt} = \xi(0) = 0. \tag{2.7}$$

Introducing the time constants τ_s and τ_l and assuming the viscous damping to be predominant,

$$\tau_s = \frac{\Theta}{\Pi} \ll \tau_l = \frac{\Pi}{\Delta}, \tag{2.8}$$

Eq. 2.5 yields to

$$\Xi(s) = \frac{\tau_s \tau_l s}{\tau_s \tau_l s^2 + \tau_l s + 1}\Omega(s) \approx \frac{\tau_s \tau_l s}{\tau_s \tau_l s^2 + \tau_l s + \tau_s s + 1}\Omega(s) = \frac{\tau_s \tau_l s}{(\tau_s s + 1)(\tau_l s + 1)}\Omega(s) \tag{2.9}$$

resulting in the transfer function of the semicircular canal G_{sc}

$$G_{sc} = \frac{\Xi(s)}{\Omega(s)} = \frac{\tau_s \tau_l s}{(\tau_s s + 1)(\tau_l s + 1)} \tag{2.10}$$

The short time constant τ_s has been found to be around $0.005s$ [37] or in a more detailed model to be close to $0.004s$ [105]. Young [145] assumed the longer time constant τ_l to be $5 - 10s$, Fernandez [38] measured $\tau_l = 5s$ for the similar vestibular organ of the squirrel monkey. A more recent study in humans estimated the time constant to be around $4s$ [24]

For typical head motion between $0.1Hz$ and $10Hz$ the bode plot (Fig. 2.4) of the semicircular canal's transfer function shows a nearly constant amplitude of the output angular rate and a phase shift close to zero.

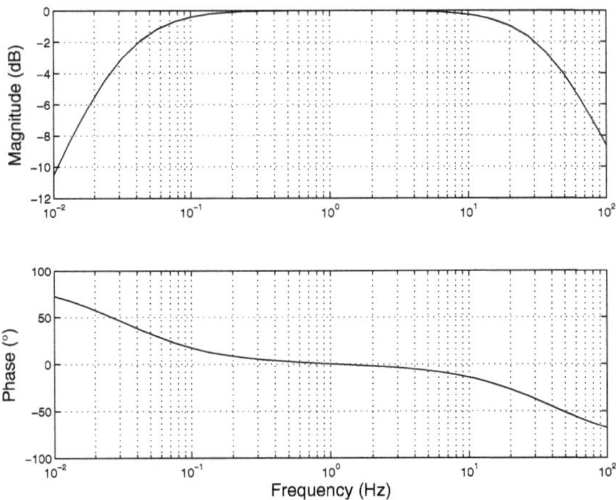

Fig. 2.4: Standardized bode plot of the dynamic behavior of a semicircular canal.

2.2 Gaze Stabilization in Biological Systems

In addition to spatial orientation, the vestibular organ is crucial for gaze stabilization in biological systems. Uncompensated head motion of only 2 - 3 °/s would already lead to a blurred image on the retina. Under pathologic conditions with impaired eye stabilization, humans would not be able to recognize faces while walking. The beating of the heart would even disturb reading while lying in bed.

Human gaze stabilization can be divided into two basic mechanisms. The optokinetic reflex/nystagmus (OKN), which mainly relies on visual information, compensates for low frequency disturbances up to approximately $1Hz$. The OKN can generate responses to full-field visual motion in order to stabilize the visual image on the retina. Monkeys and humans which possess foveal vision have developed yet another system to stabilize gaze and to follow points of interest which are moving in relation to the eye's position: the smooth pursuit system. Eye movement commands are directly generated from information derived from the visual image. For the horizontal OKN, maximum eye velocities of up to 100 °/s can be observed. The corresponding gain, which is the ratio of eye and stimulus velocity, attains about 0.8 [16]. The vertical system is similar to the horizontal, but the torsional compensation with a gain of less than 0.2 is under developed. Due to the slow "image processing" in the retina and the visual cortex, first compensatory eye movements, the so-called ocular following response, can only be observed after a latency of about $75ms$ [91]. Therefore, this system is too slow for compensating most head and body movements which peak in the range of 1 - $5Hz$ [126].

For higher frequencies, the sensory input from the vestibular organ, which has a dynamic response up to $10Hz$ (Fig. 2.4), is required. In the so-called vestibulo-ocular reflex (VOR), head motion is sensed in terms of rotation (rVOR) and translation (tVOR) by the semicircular canals and the otoliths, respectively. These signals are then used for eye stabilization. Exemplarily, emphasis in this section is put on the information flow of the rotational VOR.

The major pathway for this reflex is the well-defined trineuronal arc of the primary and secondary vestibular neurons (vn), as well as the oculo-motor neurons (mn) [76]. Due to the short latency of only $7ms$, this connection leads to a quick response to head motion. However, it is not calibrated, non-adaptive and time-invariant. With physical characteristics of the vestibular organ or the eye muscles changing with the age of a person, this direct pathway is not sufficient for perfect gaze stabilization.

There therefore exists a second adaptive pathway to cope with time-varying effects. The signals of the semicircular canals (scc) are relayed to granule cells (gc) via mossy fibers (mf). Parallel fibers (pf), which are axons of the granule cells, form one input to purkinje cells (pc) in the flocculus and carry an efference copy of the eye muscle commands and visual information in addition to vestibular signals. Each pc has approximately 200000 synapses with parallel fibers, but only one output that is relayed via the floccular target neurons (ftn) to the motor neurons. This

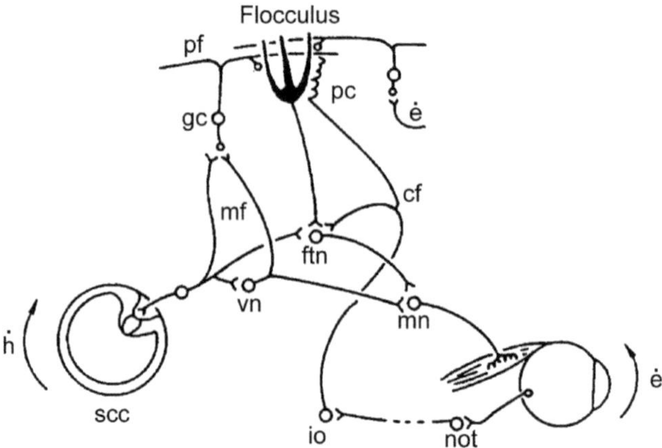

Fig. 2.5: Schematic representation of information circuits for the rotational vestibulo-ocular reflex with head velocity $\dot{h}$ and eye velocity $\dot{e}$.

cerebellar model has originally been proposed by Albus [1] and Marr [94]. Ito and others believe that adaptation takes place in the flocculus. Climbing fibers (cf) coming from the inferior olivary nucleus (io) carry information on retinal slip, which is relayed by the nucleus of the optic tract (not), as the teaching signal for the synapses of parallel fibers with purkinje cells [77]. A second input to the inferior olive comes from the brainstem and possibly carries an estimate of current eye velocity (Fig. 2.6). While simple spikes of the pc occur at a relatively high frequency of $50Hz$, the teaching complex spikes of the cf have a frequency of only $1Hz$ [76]. This general idea of inertial gaze stabilization with VOR adaptation has been extensively modeled and simulated by Glasauer [49], [51]. Fig. 2.6 shows the abstracted model of the information flow for the one-dimensional case of adaptive eye stabilization for head rotation (rVOR). In addition, Glasauer extended his model of pure inertial stabilization with smooth target pursuit in order to follow targets that are moving with respect to the eye position [52], [53]. This model, however, is non-adaptive and will be the basis for adaptation in Chapter 5.3.

Dean and Porril focused on the motor learning component of gaze stabilization [111]. By emphasizing the recurrent pathway (e.g. adaptation concerning signals containing the efference copy of eye muscle commands), they want to overcome the lack of knowledge on the inverse dynamics of the eye plant [26], [25], [65]. This idea of *cerebellar control* has been transferred to the control of robotic arms by Van der Smagt and Hirzinger [133], [134].

In contrast to the cerebellar concept of a single point of adaptation in the flocculus, there is evidence of at least a second site of plasticity. Adaptation in presence of smooth pursuit movements requires multiple sites of learning. Lisberger and others ([92], [101], [112], [14]) also believe the climbing fibers to carry the teaching signal to

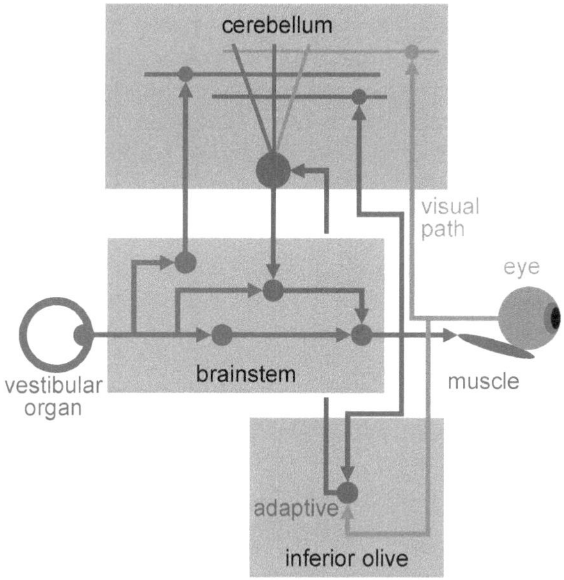

Fig. 2.6: Adaptive model with plasticity in the cerebellum.

the cerebellum, however propose additional plasticity at the floccular target neurons, Fig. 2.5.

The development of these brain models has been supported by experiments with monkeys, for example by Cullen [71].

2.3 Inertial Navigation Systems

2.3.1 Definitions

For the determination of spatial orientation two basic concepts have to be distinguished. In *gimbaled systems*, inertial sensors are mounted on a pivotable platform. The gyroscopes are used to detect potential rotation of the platform and to send stabilizing commands to its suspension frames. By moving these two or three gimbaled frames, the platform can be kept in an inertially stabilized orientation. Alternatively, spinning inertias can be used to passively keep the platform stable. For example, the orientation of a spacecraft with respect to an initial condition can be determined by measuring the joint angles of the gimbaled frame. Such a system was used for the Apollo spacecraft. The main disadvantages of this setup are its high costs and large design envelope with little potential for miniaturization. Therefore modern navigation relys on *strapdown systems*, where inertial sensors are mounted bodyfixed on a vehicle without the need for moving parts [131]. With the knowledge of its current orientation, all sensor signals can be converted from the body frame to the navigation frame. Depending on the required accuracy, this principle is suitable for low-cost miniaturization since the focus goes from hardware to software integration. Strapdown navigation systems can be structured with respect to their functional integration, Fig. 2.7. The core is a simple inertial sensor assembly (ISA) with gyroscopes, accelerometers and, if required, magnetometers, with non-processed and potentially analog output [33], [22].

In a second step, inertial measurement units (IMU) are enhanced with analog-digital-converters (ADC), a microcontroller (CPU) with some prefiltering and suitable interfaces to overlaying systems. An inertial measurement system (IMS) comprises coordinate transformation, filtering and multi-sensor fusion for orientation determination, which is discussed in Chapter 4 of this book. In miniaturized or high-end flight navigation systems, this additional processing unit is often located separately in the vehicle [72].

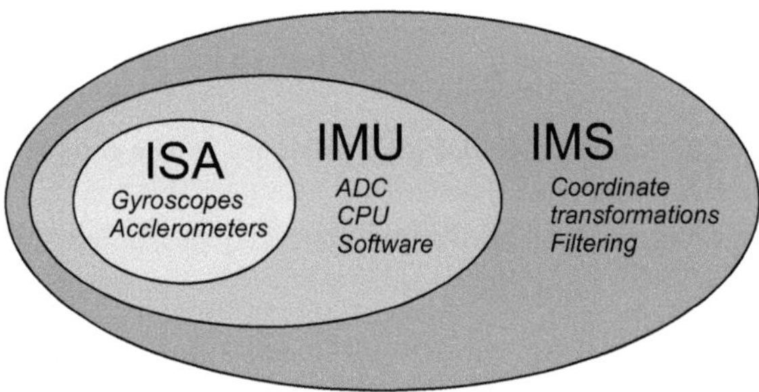

Fig. 2.7: Categorization of inertial navigation systems.

2.3.2 Sensor Components

As mentioned above, inertial navigation systems always consist of at least gyroscopes and accelerometers. In case heading is required to be calculated, IMS are mostly enhanced using an additional magnetometer, thus resulting in a new setup called Attitude and Heading Reference System (AHRS). This is needed since with gyroscopes and accelerometers alone, a rotation around the vector of gravity cannot be time-stabilized and sensor drift would lead to error for longer lasting measurements. This can occur after one minute for low-cost MEMS gyros or after several hours with high-end fiber optic gyros. But in any case, long-time stabilization of the heading angle is necessary.

Recently this problem was solved in a different way. Since measurement of the Earth's Magnetic Field (EMF) is sensitive to disturbances from peripheral activity, for example laboratory equipment, motors in the vicinity of the sensor, or by disturbances in the Earth's atmosphere, the heading angle can be time-stabilized by a two-antennae GPS system [130]. With real-time kinematic differential GPS, the position of each antenna can be determined with an accuracy of a few centimeters. Provided an antenna distance of about 4m (e.g. on a vehicle), heading can be stabilized with an accuracy of about 0.5°. In the following section, the state-of-the-art of currently used sensor technology shall be detailed.

Gyroscope Technology

Historically, gyroscopic systems use the inertial properties of a wheel or rotor spinning at high speed. Such a system tends to maintain the direction of its angular momentum and is thus inertially stabilized in space. In a gimbaled reference frame, rotation of a body to which it is attached can be measured. Since the importance of this technology is decreasing and it is not used for this book, please refer to [131] for further details. Modern commercially available high-end inertial measurement units rely on fiber optic gyroscopes (FOG) [117]. Their basic principle of operation is based on the Sagnac effect. Light from a source is split into two beams and travel

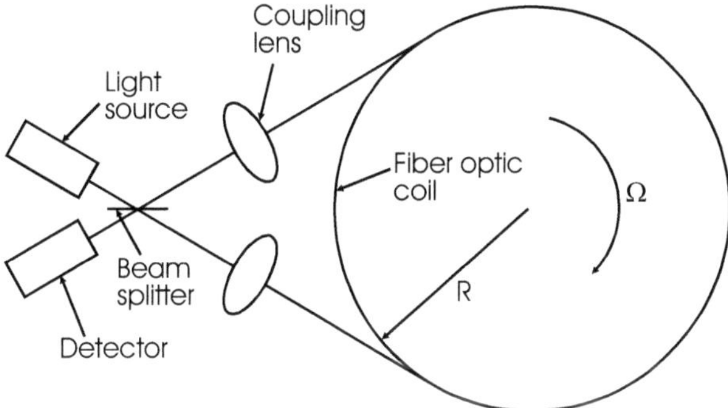

Fig. 2.8: Fundamental optical components of fiber optic gyroscope.

in opposite direction of a fiber coil. The number of turns in the coil is highly relevant since it significantly increases the sensitivity and accuracy of the gyroscope. The two light beams are then reunified and propagated to a detector which measures the phase difference $\Delta\Phi$ between the two beams. When the coil does not turn around its axis, the two beams travel the same distance and no phase shift occurs. However when the setup turns clockwise (Fig. 2.8), the clockwise beam has a higher optical path length ΔL until it reaches the detector.

The resulting phase shift

$$\Delta\Phi = 2\pi\frac{\Delta L}{\lambda} \tag{2.11}$$

with light wave length λ, velocity of light c, coil radius R and

$$\Delta L = \frac{4\pi R^2 \Omega}{c}N \tag{2.12}$$

shows to be proportional to the angular rate Ω and the number of turns N in the coil:

$$\Delta\Phi = \frac{8\pi^2 R^2 \Omega N}{c\lambda} \tag{2.13}$$

Due to the high accuracy at the relatively high cost of several thousand euros per axis, this technology is mainly used for navigation purposes in commercial aerial vehicles or for military applications.

Consumer, as well as automotive applications mainly rely on MEMS (Micro-Electro-Mechanical-Systems) technology which combines the advantages of low cost with extremely small size. During manufacturing, a silicon wafer is grown on a carrier substrate in a process called chemical vapor deposition. When the epitaxial layer has reached the desired thickness, a photo resist is deposited on it. With a light or X-Ray beam through a mask, the photo resist is removed at locations where then the unprotected silicon is etched away, thus leaving a three dimensional structure on the carrier substrate. By repeating these steps, complex multi-layer structures can be manufactured. The sensing structure of a MEMS gyroscope accumulates to about $1mm \times 2mm$, Fig. 2.9 (right).

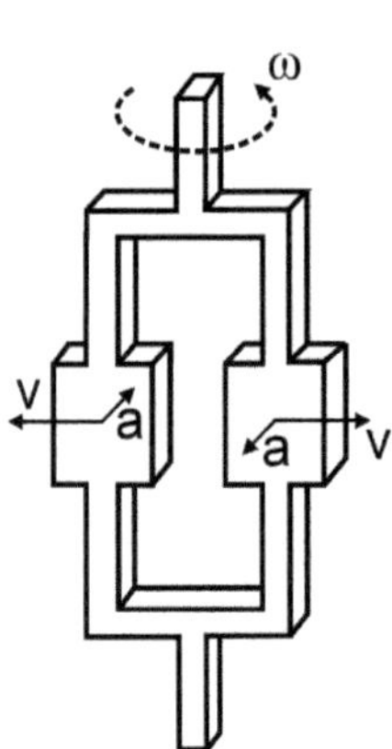

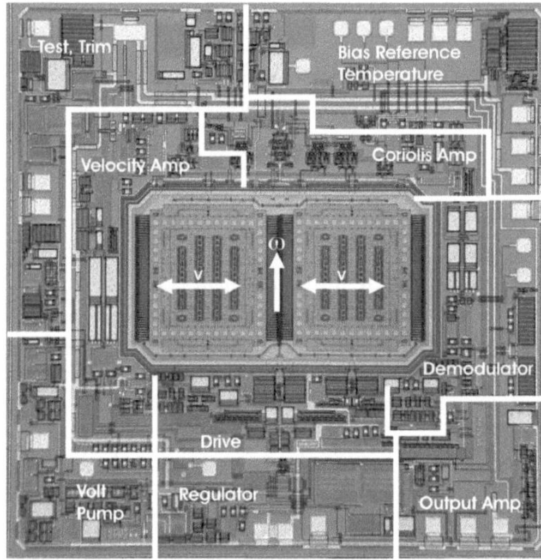

Fig. 2.9: Operating principle of tuning fork MEMS gyroscopes (*left*) and photograph of a sensor die including proof masses and signal processing circuits [45](*right*).

Fig. 2.9 (*left*) shows the basic principle of operation of acceleration compensated tuning fork gyroscopes. Two proof masses m oscillate in opposite direction. When subject to rotation around a perpendicular axis, this generates a coriolis force F_c

$$F_c = ma_c = -2m(\omega \times v) \tag{2.14}$$

with coriolis acceleration a_c, angular rate ω and mass velocity v. The coriolis force causes a displacement of the proof masses which is sensed with a capacitive comb structure. Due to the opposite oscillation of the masses, additional accelerations imposed on the structure are compensated for. By integrating the signal processing circuitry onto a monolithic IC, the market leader Analog Devices Inc. managed to further reduce the manufacturing costs of its iMEMS gyroscopes [43]. Since the process is suitable for mass production, MEMS gyroscopes are available at an acceptable price for consumer applications [44].

In addition to the limited accuracy, the major drawback of commercial MEMS components is the limited bandwidth. Up until now, bandwidths over $80 - 100Hz$ were not available.

Accelerometer Technology

Commercially available accelerometers are based on the use of Newton's principle of an inertial force acting on a proof mass. In piezo sensors, the necessary reaction force is produced by a piezo element which outputs a force-proportional voltage. In contrast to that, MEMS sensors are displacement oriented. In the presence of only little damping to stabilize the system (Fig. 2.10), the inertial force $F_a = ma$ is equal to the spring reaction force $F_c = cx$, thus resulting in a proportional relationship between acceleration a and displacement x. Due to the integrated structure of MEMS technology, layouts for the measurement of three-dimensional accelerations can be realized with a single proof mass.

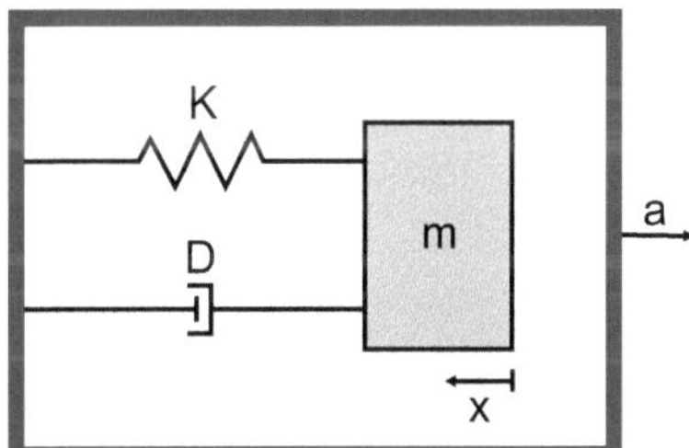

Fig. 2.10: Operating principle of MEMS accelerometers.

One possible design shown in Fig. 2.11 has a quadruple suspension of the proof mass. In neutral position, e.g. when not exposed to acceleration, the four capacitors C have equal distances between their plates, resulting in equal voltages V when subject to the same electric charge Q:

$$V = \frac{Q}{C} \approx \frac{Qd}{\epsilon A}, \quad A >> d^2 \tag{2.15}$$

with capacitor plate area A, dielectric constant ϵ and plate distance d. The electric charge is always kept constant on the capacitors. When exposed to acceleration in out-of-plane direction, the distance between the capacitor plates varies by Δd, resulting in a common symmetrical voltage change

$$\Delta U = Q\frac{\Delta d}{\epsilon A} \tag{2.16}$$

on each capacitor. For accelerations in in-plain direction, the proof mass displaces asymmetrically, thus leading to different voltage changes. This allows for the measurement of all three accelerations with one single sensor die.

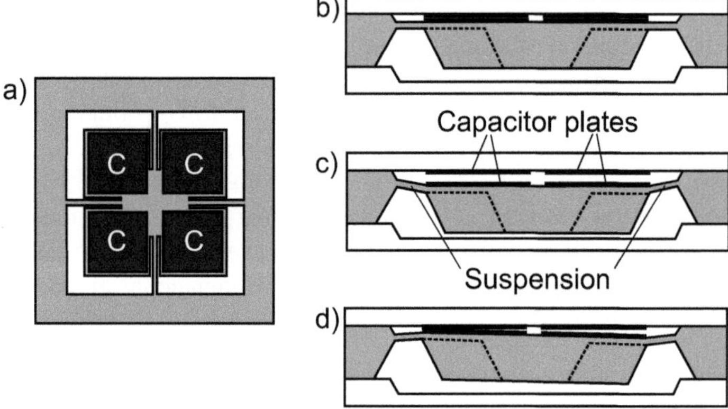

Fig. 2.11: Tri-axial accelerometer in top view (a) with four capacitors C, side view of the neutral position (b), when accelerated upwards (c) and to the right (d).

In comparison to piezo accelerometers, MEMS sensors have the important advantage not only of measuring frequencies up to several hundred Hertz but also of being applicable for constant accelerations [46]. This allows for the sensing of gravity which is necessary for most navigation purposes. MEMS accelerometers became cheap with mass production due to their application for air-bags in the automotive industry.

Magnetometer

For the determination of a vehicle's heading (e.g. rotation around the vector of gravity), inertial measurement systems mainly rely on magnetometers which are meant to measure the Earth's magnetic field. Even though this weak field of only 0.6 Gauss is easily disturbed by ferromagnetic components in the sensor's surrounding, it is suitable for stabilizing sensor fusion algorithms for drift around the vector of gravity. Today's electronic compasses can make use of different sensing principles like fluxgate or magnetoresistive sensing. Fluxgate sensors consist of a set of coils with an excitation circuitry that is capable of measuring magnetic fields with a resolution of less than 1 milligauss. The drawbacks of their bulky design envelope and their long reading response time of several seconds make them useless for miniaturized inertial measurement systems [20].

Magnetoresistive sensors are made of thin strips of Permalloy, a $NiFe$ magnetic film, whose electrical resistance varies with the change in the applied magnetic field. With four sensing strips per axis, a Wheatstone bridge amplifies the changes in resistance and generates an output voltage. Due to the low changes in the magnetic field, the output has to be amplified by a factor of the magnitude 10^3. A tri-axial magnetometer can be assembled in a 16-pin LCC package with the dimensions of $7.4mm \times 7.4mm \times 2.5mm$ [69].

2.3.3 Commercially available units

Commercially available inertial measurement units can generally be divided into two classes. For high-end navigation purposes, specialized companies offer complete systems for aeronautical and military purposes. The state-of-the-art for civil applications is AEROcontrol by IGI Systems which combines an inertial system based on fiber optic gyroscopes with an integrated GPS receiver and a proprietary control unit, Fig. 2.12 (*right*). These systems achieve an angular accuracy for roll and pitch of 0.004°and 0.01°for heading [73]. AEROcontrol can achieve an update rate of 256Hz. However, with a weight of more than 8kg and a price of 50,000 euros IGI Systems are not suitable for portable or consumer related applications.

Inertial sensors by iMAR GmbH are adequate for a wider use. The iVRU-FAS series units (Fig. 2.12, *left*) with fiber optic gyroscopes and integrated strapdown processor determines roll and pitch attitude at an accuracy of 0.1°[74]. Heading information must be provided from outside. With a cubical size of about $120mm$ and a weight starting at 1.6kg, these systems are better suited for portable applications like on biped robots. The price of 20,000 euros is the major drawback.

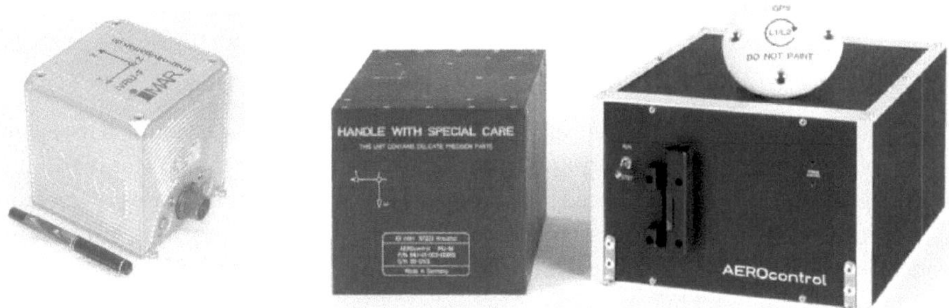

Fig. 2.12: Inertial measurement unit with fiber optic gyroscopes by iMAR GmbH (*left*). AEROcontrol IMU and processing unit with GPS antenna (*right*).

The second main group are inertial measurement units completely based on MEMS technology. Due to the components' low price and design envelope, these sensors find a wide application in automotive and robotic applications as well as for consumer electronics. Since the MEMS fabrication process is a mass production of silicon wafers in a critical clean-room environment, only few companies produce the sensor dies themselves. Most IMU manufacturer are so-called value-added resellers for sensor components, mainly supplied by Analog Devices, Inc. The smallest available IMU is currently commercialized by Cloud Cap Technology, Inc. and has a design envelope of $34mm \times 38mm \times 20mm$ without housing. The sensorhead alone, an ISA with additional ADC, is available at $28mm \times 29mm \times 15mm$ [21].

The second major MEMS-IMU manufacturer Xsens Technologies B.V. focuses on the application of its sensors. The main product MTi is an AHRS consisting of gyroscopes, accelerometers and magnetometers. In a full body Lycra stretch suit, the manufacturer integrated 16 MTi sensors on each foot, leg part, arm part and hand as well as on the trunk and the head. A proprietary full body kinematic model can then be animated and evaluated in real-time with the sensor data. The setup is able to compensate for electromagnetic disturbances up to 30s [144].

In between these two main groups are several niche market suppliers which typically use carefully selected and individually tested standard MEMS components or slight technological modifications. Due to the wide range in production tolerances, a small percentage of standard MEMS sensors can even reach MIL-grade performance characteristics. This selection process is highly expensive and results in a price of about 7000 euros for those IMUs (Systron Donner Inertial, MMQ50 [127]).

2.4 Active Camera Stabilization in Technical Systems

Camera stabilization techniques are best known from photography. Small digital cameras usually take a larger image than required, and then sense camera rotation and shift the final picture frame within the larger image. This can help to reduce low frequency disturbances if the image quality is already good, but does not improve the output if the acquired image itself is blurred.

In that case a stabilization of the optical path is required. Professional lenses, like those of photo journalists, usually incorporate two gyroscopes for pan and tilt motion. Since these systems only need to compensate for small amplitudes of motion, they do not need electric motors that provide high angles of rotation, but rely on bi-concave lenses that are subject to in-plane shifts with miniaturized actuators [19]. Fig. 2.13 shows the working principle.

An image stabilizer lens is located in between the focusing group and some other lens groups. When the camera is accidentally pointing down due to a trembling

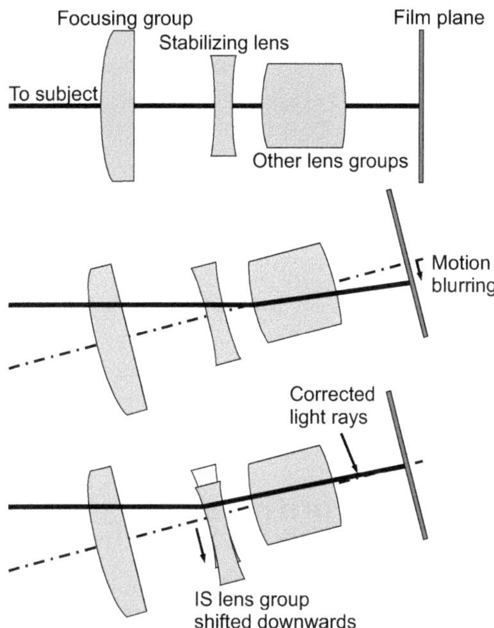

Fig. 2.13: Image Stabilizer in professional photo lenses.

photographer, the center of the original image would move downwards, as shown in Fig. 2.13 . To compensate for that, the bi-concave lens is shifted down to refract the center light beam at a spot with different inclination to correct the light rays. This system does not completely compensate for all motion but increases the maximum focal length that can be used [17], [18].

There exist several attempts to derive camera stabilization in technical systems from neuroscientific models. Shibata, Schaal and coworkers developed a model of stabilization and smooth pursuit based on online learning of the forward model of target motion (e.g. predicting pursuit commands [122]). Their model of biomimetic gaze stabilization is based on feedback-error-learning with regression networks. As a teaching, but also as a direct compensatory input, retinal slip is used. In the adaptive version, a static (e.g. not moving target) is assumed [121]. Adaptation has been proven in simulation at a sinusoidal 0.5Hz motion and on a humanoid head setup at 0.8Hz. Work by Kettner et al. goes into a similar direction [86].

Panerai and Sandini focus on robotic applications, especially for humanoids. Their main contribution is the focus on integration of inertial and visual information [109]. In a binocular system with realistic inter-ocular distance, target position is calculated and serves for improving the visual command input. A neural network also uses residual optic flow as a teaching signal and optimization criteria [108]. Experiments prove good performance in absence of target self-motion [107], [106].

In aeronautics, stabilized cameras became popular for surveillance helicopters. Federal border patrol teams use infrared cameras located under the helicopter to detect people in darkness. The police uses these systems for vehicle tracking and recognition. Some aircraft have similar setups for the inertial stabilization of radar antennas.

The automotive industry has just begun commercializing car-fixed infrared wide-angle cameras for obstacle detection by night. Pivotable cameras are still subject to research and development. Pioneering work has been pursued by Dickmanns that culminated in an autonomous drive from Munich to Denmark and back. Vision for Dickmann's vehicles was achieved by the so-called Multi-focal Active/Reactive Vehicle Eye (MarVEye) which consists in some versions of pivotable yaw-platform with an active mirror for gaze stabilization in tilt-direction [30].

2.5 Sensor Fusion Algorithms

The basic problem of orientation calculation is the inadequacy of individual sensors and technologies. Each kind of sensor has its own properties, characteristics and strengths. Gyroscopes deliver good short-term information, but if integrated over a long time, drift is unavoidable. Linear accelerations are not subject to drift if not integrated, however orientation statements can be made only with respect to the vector of gravity. Overlaying accelerations (e.g. due to shocks) significantly disturb the accuracy of orientation calculation. The same is true for magnetometers with respect to the Earth's magnetic field.

Thus results the necessity for sensor fusion algorithms that combine the best information from all sensors at specific working conditions.

Except for the use of magnetic data, the biological system of orientation perception is similar to the technical approaches and uses accelerations and angular rates of the head [7], [99]. The main problem is also sensory equivalence between gravity and other accelerations [66], [54]. Extensive modeling of self-motion perception has been done by Holly [67] [68], Zupan [148] Raymond [113], Merfeld [148] and Glasauer [54].

Technical approaches similarly combine information from all available sensory sources. They tend to use gyroscope data for short-term accuracy and acceleration as well as magnetic field readings for long term stability. A common approach for multi-sensor fusion is the Kalman Filter (KF), which is a solution of the Wiener Problem of noisy random processes without the need of infinite memory since it does not store all past data [82]. Good engineering applications of this highly theoretical and mathematical work are given by Maybeck [98] and Brown [15].

Standard Kalman Filters are valid for linear process models only, which restrict inertial data fusion to low amplitude motion. Implementations are documented by

Dissanayake [31], Ang [6] and many others. If low amplitudes can be assumed and process as well as measurement noise have constant covariances, a simplified method introduced by Baerveldt [10] as complimentary filtering can be derived. It is based on low-pass filtering of accelerations and high-pass filtering of gyroscopes. This computationally inexpensive method yields good results for the humanoid robot JOHNNIE, which keeps his upper body in an upright position [47] [48].

If motion cannot be reduced to low amplitudes transformation matrices become nonlinear. For those cases Extended Kalman Filters (EKF) make a first order approximation of the transition matrix to better cope with nonlinearities [139]. To overcome the limitation of linearization, the unscented transformation (UT) was developed as a method to propagate mean and covariance information through nonlinear transformations, which yield better results in highly nonlinear processes [78], [79]. Sigma-point Kalman Filters (SPKF) make use of the UT for sensor data fusion in inertial navigation [132], [110], [100]. They are especially suitable for tightly coupled GPS/INS integration [140], [141], [146]. SPKF have attracted attention during the last 2-3 years and need further research and evaluation.

3 Development of a Highly Miniaturized Inertial Measurement System

Within the project FORBIAS, it was necessary to develop a completely new inertial measurement system. The main reason for not using a commercially available system was the relatively large design envelope that is contradictory to the goals of the project. In addition, most units specify the interface protocol, but do not allow the user to directly alter and adapt the IMS's firmware. Therefore, the local microcontroller could not be used for customized signal conditioning algorithms. This chapter will detail the development of μCube, a highly miniaturized inertial measurement system.

3.1 Requirements Analysis

The main goal of the new IMS was to fulfill the varying requirements of the different applications within the FORBIAS project, from driver assistance (A2), to helicopter pilot assistance together with EUROCOPTER, to the specific needs of the FORBIAS head-mounted camera (B2) [40]. In addition, the IMU should be an independent and market-ready product. Consequently, from this originated several requiremens for the new inertial measurement system that cover a wide variety of different applications:

- **Physical properties**
  The design envelope is the most critical requirement for the new IMS, since it is the main reason for not using commercially available systems. In order to allow for a reasonable use with the head-mounted camera, a cubical design with an edge length of 20mm is the maximum accepted size. In addition, the camera manufacturer (A2) limited the overall mass to 15 grams. For the automotive use with driver assistance systems, size is most critical since in a final application, the IMS should fit into the rear-view mirror on the windscreen of a car. Therefore an edge length of less than 20mm is appreciated by the involved car manufacturer and is a unique selling point for the product.

- **Sensor components**
  To fulfill the project requirements for gaze stabilization and orientation calculation, tri-axial sensing of angular rates with gyroscopes and of linear accelerations is required. Due to the priority of size and the primary FORBIAS

W. Günthner: *Enhancing Cognitive Assistance Systems with Inertial Measurement Units*, Studies in Computational Intelligence (SCI) **105**, 25–48 (2008)
www.springerlink.com

applications, magnetic field sensing should not be integrated into the main IMU but the IMS should provide a suitable interface to connect a magnetometer. Since manufacturing costs are crucial for OEMs, only standard electronic components should be integrated. However, the minimum temperature range from $-40°$C to $+80°$C must be attained. An integrated temperature sensor for the compensation of thermal influences would be an additional feature, but is not mandatory.

- **Measurement range, resolution and performance**
  The measurement range for the IMS is determined by the operating conditions in the automotive and aerospace application, as well as by the head dynamics for the head-mounted camera. Accelerations in the horizontal plane of vehicles are limited by the friction between wheels and the road to about 1g. In vertical directions bumps can cause accelerations of a little higher magnitude. For regular driving without misuse, BMW specifies accelerations of up to 1.2g. Angular rates usually do not exceed 60 °/s. With the first eigenfrequency of a car chassis of about 2Hz, the dynamic sensor response is not a factor. Due to the capabilities of the BMW driving simulator, frequencies of up to 10Hz should be measureable [39]. For the helicopter application, specifications could not be determined before test flights. It was only known that angular rates do not exceed 45 °/s and that the first eigenfrequency is about 26Hz. The head-mounted application of the inertial measurement unit requires angular rates of up to 300 °/s [40]. Accelerations and dynamic responses are negligible in comparison to the other applications.

  The two different scenarios need to be distinguished. The automotive and aerospace applications require a high resolution of better than 1mg in linear acceleration and 0.1 °/s in angular rates. For the head camera, 0.25 °/s is sufficient, since angular rates are usually higher.

  Due to the rather low frequencies to be resolved, a high output rate of the inertial measurement data would not be required. However, the frame rate directly influences latencies between motion initiation and the first corresponding sensory output. Therefore, a frame rate of 1kHz is required to keep maximum latencies lower than 1ms, which is reasonable in comparison to the equivalent biological system.

- **Electronic requirements**
  The new inertial measurement system must be suitable for in-field measurements with either not appropriate or not voltage stabilized power supplies. Therefore, it needs a wide range of available supply voltage and has to work with a standard 9 Volt battery.

- **Software requirements**
  The microcontroller software is crucial for the performance of the whole inertial measurement system. Real-time filtering algorithms for sensor noise reduction

have to be implemented in order to perform primary signal conditioning decentralized on the IMS. Inertial measurement units operate in a great variety of environments with different requirements. Therefore bus protocol, frame rate, and other parameters must be changeable in-field by the operator. In intelligent sensor-actor-networks, individual components often need to be time synchronized. The new IMS shall provide an input for external triggering via one of its standardized interfaces.

- **Interfaces**
  Varying applications within the FORBIAS project impose the need for two different interfaces. In the automotive and helicopter application, a 1Mbit/s CAN2.0A bus should assure high transmission rates and compatibility with automotive standards. Control algorithms for the head-mounted camera, however, will run on a notebook, so an additional RS232 interface is required for easy connectability with standard PCs.

3.2 Sensor Component

Sensor components are the heart of μCube. The following sections focus on the choice of gyroscopes and accelerometers. In addition, magnetic field sensors are mentioned although they are not part of the μCube, but external components, since they are necessary for the orientation determination in Chapter 4.

3.2.1 Angular Rate Sensors

The choice of the gyroscopes is crucial for the performance of the IMS since they influence especially short term errors that are more difficult to compensate for. Except for the Murata Gyrostar series sensor, which takes advantage of the piezoelectric principle, all gyroscopes that are considered for the μCube, due to their small size, are based on the MEMS technology as described in Chapter 2.3.2. On the sensor market, only single axis gyroscopes are available.

Analog Devices Inc. (ADI), world market leader for MEMS inertial sensors, distributes two gyroscopes based on the same die but with different measurement ranges. The ADXRS150 with high sensitivity is especially suitable for the automotive and aerospace application, whereas the ADXRS300 provides the necessary measurement range for the head-mounted camera. Noise is the most critical characteristic for inertial sensors. Nonlinearities, bias drift vs. temperature etc. can be identified and compensated for, however, noise is a random parameter that can only be filtered to a certain extent. Therefore ADI gyroscopes are superior to the other choices. Both ADI chips are pin-compatible and can be used with the same printed circuit board (PCB). This makes the series ideal for the different FORBIAS applications. The

SAR10 is not suitable since its 10bit SPI interface does not provide the necessary resolution, Table 3.1.

Table 3.1: Characteristics of relevant gyroscope dies ([5], [4], [120], [102], [12]).

Manufacturer	ADI	ADI	Sensonor	Murata	Bosch
Type	ADXRS300	ADXRS150	SAR10	ENC-03J	SMG061
Range [°/s]	300	150	250	300	240
Sensitivity [mV/(°/s)]	5	12,5	SPI	0,67	7
Nonlinearity [%FS]	0,10	0,10	0,30	5,00	0,50
Bandwidth [Hz]	80	80	50	50	30
Noise [°/s rms]	0,9 @80Hz	0,45 @80Hz	1,5	N/A	1,5
Price [$]	51	51	N/A	40	N/A

For the further development of the μCube, the ADXRS series by Analog Devices, Inc. has been chosen.

3.2.2 Accelerometers

For the calculation of spatial orientation, a reliable determination of the vector of gravity is needed. Since piezo sensors only measure dynamic accelerations, gravity does not generate an output signal. As seen in Fig. 2.11, MEMS sensors do not have this drawback. Table 3.2 shows a selection of relevant accelerometers. The integration of three sensitive axes into one chip, like in the ST and Bosch sensor, is a clear advantage for miniaturization.

Table 3.2: Characteristics of relevant accelerometers ([125], [3], [13], [124]).

Manufacturer	ST Microel.	ADI	Bosch	Silicon Design
Type	LIS3LV02DQ	ADXL103/203	SMB363	1221
No. of Axes	3	1/2	3	1
Range [g]	2/6	1,7	2	2
Sensitivity	1 mg/LSB	1000mV/g	600mV/g	2000mV/g
Nonliearity [%FS]	2	0,5	0,5	0,3
Bandwidth [Hz]	640	2500	1000	400
Noise [$\mu g/\sqrt{Hz}$]	8	110	200	5
Price [$]	16	8/12	N/A	203

Originally the ADXL series was used until the LIS3LV02DQ became commercially available. The outstanding noise performance, equivalent to a triaxial sensory system at more than $600 (Silicon Design), led to favoring this ST sensor for the final design.

Having an all digital output with SPI interface, which requires no further analog filtering, eases connections and signal routing on the PCB.

3.2.3 Magnetic Field Sensors

For orientation determination including heading angle, the μCube is extended with a magnetic field sensor. The chosen HMC1053 from Honeywell SSEC has a field range of $\pm 6gauss$, which is wide enough to cover disruptive laboratory influences and small enough to resolve the Earth's magnetic field (EMF) sufficiently.

The three analog outputs are filtered and amplified by a factor of 600 to 1000 in order to use the full range of analog-digital conversion. With a supply voltage of $5V$, the EMF only causes an output signal of about $\pm 3mV$. To keep sensor sensitivity high, magnetometers like the HMC1053 regularly need set/reset cycles with $4A$ current impulses for several milliseconds in each direction, resulting in superimposed magnetic fields of more than $40gauss$. These fields realign the magnetic domains in the magneto-resistive permalloy film strips [70]. The unknown and time-varying sensitivity of the magnetometer needs special care in the control algorithm that is using the data.

Fig. 3.1 shows all sensory components which are used for the μCube and magnetic field measurement.

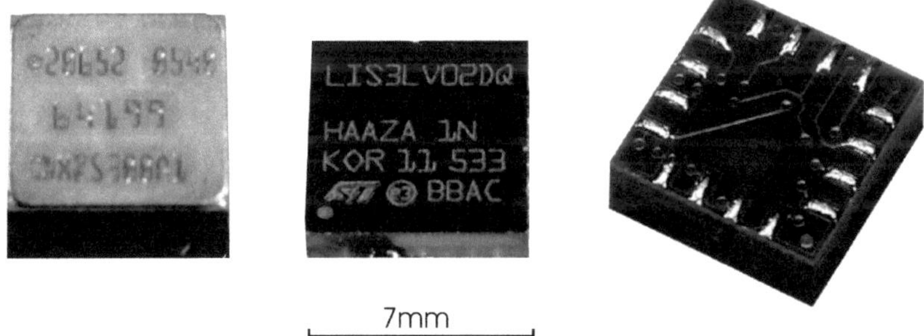

Fig. 3.1: The three used sensorcomponents: gyroscope ADXRS300 (*left*), accelerometer LIS3LV02DQ (*middle*) and magnetometer HMC1053 (*right*).

3.3 Electronic Concept and Microcontroller

The new inertial measurement unit μCube is designed as an element of an intelligent sensor-actor-network, requiring an onboard CPU for the execution of device control software, as well as the necessary filtering algorithms.

Fig. 3.2 visualizes the hierarchical electronic structure of the unit. The μCube is controlled by an 8bit Atmel AVR AT90CAN128. With 128kB of flash and 4kB of EEPROM memory, it is capable of storing and executing the software code. The main advantage is its onboard CAN controller, which makes the usually needed external chip obsolete, saving the $18mm \times 10mm$ PCB footprint that the standard SJA1000 would need. Simple programming in C language helps the end user implement proprietary code. The microcontroller is driven by an external 16MHz quartz to ensure the required cycle accuracy for high-speed CAN and to reduce transmission time on the onboard SPI bus.

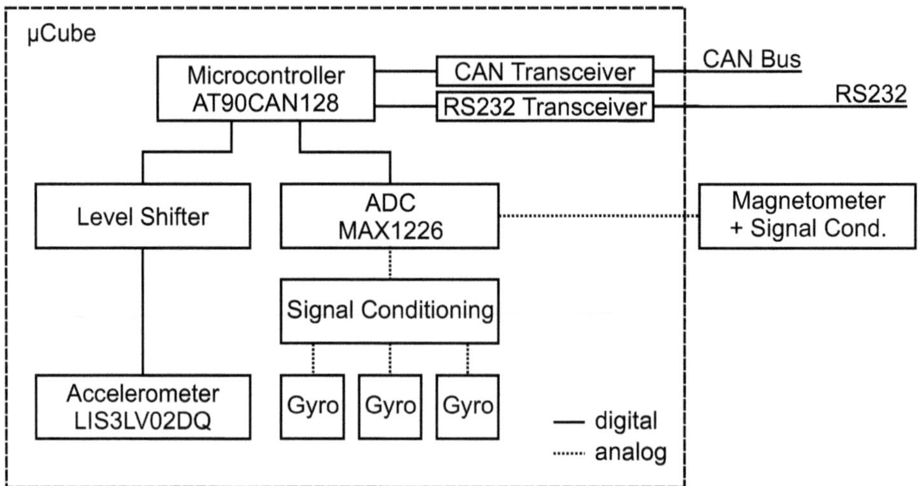

Fig. 3.2: Electronic setup of the μCube and periphery.

Sensor data acquisition is done in two different ways. The triaxial accelerometer provides a digital SPI interface for a direct connection to the CPU. Since the accelerometer operates on 2,5V and the microcontroller on 5V, level shifting of the four signals clock cycle, device selection, and data-in and data-out is mandatory.

Also via SPI bus, the Analog-Digital-Converter MAX1226 is connected to the CPU. Before receiving the analog signals from the gyroscopes and possibly magnetometers, signal conditioning, e.g. low-pass filtering is required to prevent aliasing effects.

For communication with a superordinate master control and with other sensors or actuators, the μCube provides two digital external interfaces. On a CAN bus, it can interact directly with a high number of receivers in parallel. Due to the high transmission rate of 1Mbit/s, it is especially suitable for low-latency applications. On the other hand, the usually point-to-point RS232 interface provides serial connection to standard receivers like PCs or FPGAs. The chosen RS232 transceiver allows for transmissions with up to 400Mbit/s. Serial connections are also used for updating the IMUs firmware, as will be explained in Section 3.4.

In a highly integrated analog-digital environment, special attention must be given to power supply sources. Influences from the digital circuits of the microcontroller and the interface transceivers have severe effects on the sensor noise performance. Therefore, the digital part must be separated structurally and in layout as far as possible. A special status has the accelerometer, which requires supply for its analog sensing part and uses the same source for its digital SPI interface. To block backpropagating effects, a separate voltage regulator is used, Fig. 3.3.

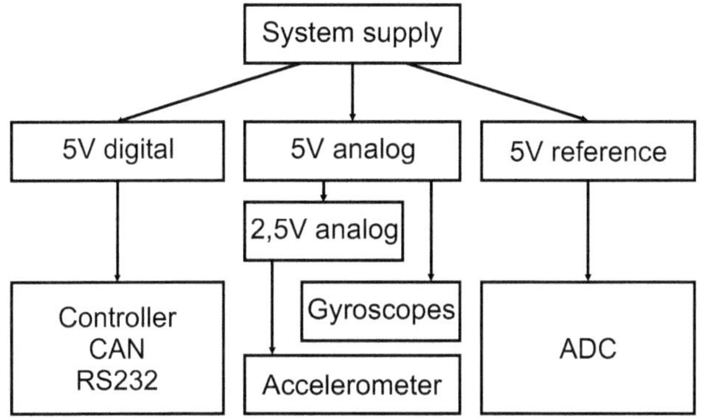

Fig. 3.3: Power concept for μCube.

The whole system can be operated with supply voltages between 7V and 10V, which do not need to be stabilized. This makes μCube suitable for in-field operation with, for example battery supply.

3.4 Software

Microcontroller software is one of the critical aspects of the device design. There is no alternative to real-time task execution, which leads to an integrated and interrupt driven software concept. AVR microcontrollers are structurally well suited for optimized programming in C language, e.g. efficiency losses in comparison to machine coding are less significant than for other cores. Therefore, most of the software is programmed in C. Only highly time-critical parts are coded directly in assembly machine language.

The necessary flexibility, as stated in the requirements analysis in section 3.1 is ensured by a hierarchic software structure. During manufacturing, a bootloader is programmed in the EEPROM of the CPU with the *in system programming* (ISP) protocol and cannot be changed later since the involved microcontroller pins are used for different purposes afterwards. At every power-up cycle, this bootloader is executed and waits 10 seconds for contact with a host PC to receive new firmware. After an update, or if no contact is established, the firmware itself – which contains the actually intended software – is executed. In the beginning, the hardware components of μCube are initialized. This includes for example measurement range and bandwidth settings of the accelerometer, as well as the initiation of communication via the external interfaces CAN or RS232.

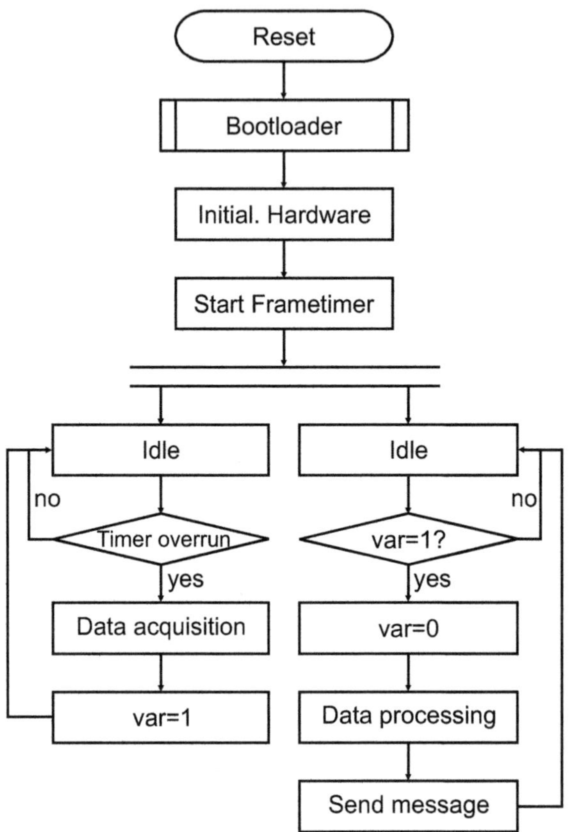

Fig. 3.4: Firmware of the μCube.

Fig. 3.4 shows the flowchart of the microcontroller software. The main part consists of two separate routines, one responsible for data acquisition, the other for processing and sending. Timer interrupt driven, the microcontroller fetches new measurement data directly from the accelerometer via SPI and from the ADC which converts gyroscope and possibly magnetometer signals. At the same time, a new A/D conversion cycle is started. When all new measurements are available in the CPU, the data processing algorithm is initiated. Then the sensor values are sent via one of the external interfaces and the microcontroller goes into idle mode until the cycle is over. Since the tasks are interrupt-driven and especially the CAN bus is non-deterministic, only a typical cycle can be investigated. Fig. 3.5 shows a timing analysis for a μCube with $1kHz$ frame rate, e.g. a cycle time of $1000\mu s$ and an accelerometer/gyroscope setup using the $1Mbit/s$ CAN interface unidirectionally to a superordinate master control.

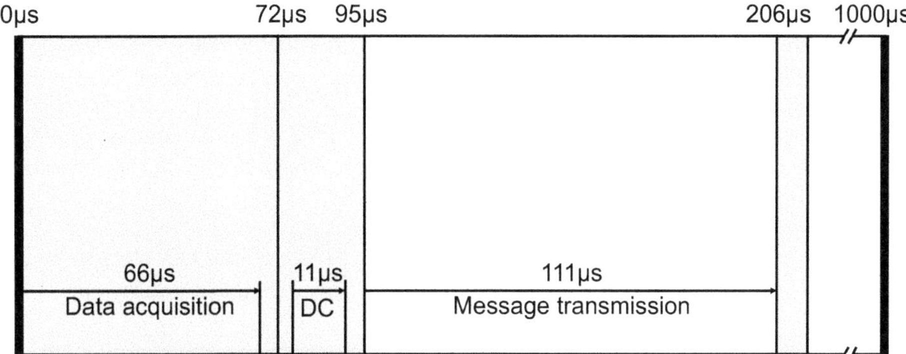

Fig. 3.5: Timing analysis of the microcontroller firmware, with the periods of active CPU usage shaded grey.

For the timing simulation the timer overrun interrupt service routine (ISR) is chosen as a deterministic cycle beginning. After $66\mu s$ sensor data from accelerometers and gyroscopes are available in the microcontroller, the new A/D conversion is started. At $72\mu s$ the data processing routine is started. Its main content is a loss-free compression algorithm to fit the 6 times $12bit$ sensor data into the $64bit$ data frame of a CAN message, Section 3.5. The transmission of the CAN message itself takes $111\mu s$, followed by $8\mu s$ in which the CAN controller interrupt done routine is executed. Thus, at a cycle time of $1ms$, the microcontroller is in idle mode for $897\mu s$ [2].

The limiting factor for higher frame rates without loss of quality is A/D conversion. One conversion typically requires $4.1\mu s$ [97], depending on the ADC temperature and external clocking. With $32\times$ oversampling (Section 3.7) and three gyroscopes, pure conversion takes $396\mu s$. Including the $66\mu s$ for data acquisition, multiplied with a security factor of 1.2, the minimum cycle time adds up to $555\mu s$, leading to a maximum frame rate of $1.8kHz$. In case of unidirectional use of the CAN bus or accurate bus scheduling, this is the maximum frame rate μCube can attain without loss of data quality.

3.5 Interfaces

As sketched in Fig. 3.2, two digital external interfaces are available on μCube. Especially in automotive applications, CAN bus is the standard technology for the interconnection of peripheral devices. Car manufacturers and their suppliers use medium speed (250-500kBit/s) CAN to control and connect electronic subsystems. If a private bus on the vehicle is available, μCube uses high-speed 1Mbit/s CAN for reduced transmission times and low latency, but if it has to be connected to an existing bus, the rate can be lowered to any other standard value.

One single CAN2.0A message can have a length of up to $108bit$, consisting of an 11bit identifier (CAN ID) and a data field of $0 - 8$ $Byte$, e.g. each message carries $44bit$, or more than 40%, overhead. The required $72bit$ for all six sensor values of $12bit$ each, would lead to two messages per frame. This is highly undesirable since each message accounts for more than 10% bus usage, with 50% being the reasonable limit for non-deterministic bus arbitration.

Therefore, a loss-free data compression algorithm (DC) is implemented to reduce CAN bus load by more than 30%. Fig. 3.6 details the message composition.

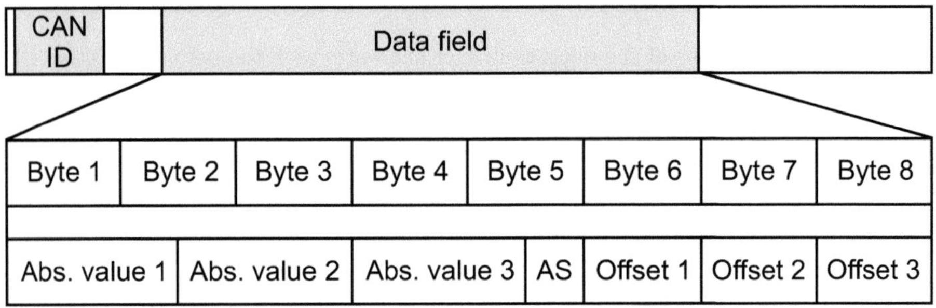

Fig. 3.6: CAN message structure in compressed mode and with CAN standard identifier.

Two alternating messages with different IDs and content are sent. For a frame rate of 1kHz, for example, message A is followed by message B $1ms$ later, followed by A another millisecond later. Message A transmits the absolute values of the three gyroscopes in the first $4.5Bytes$. The second part contains the offset of the new accelerometer reading including the algebraic sign (AS) with respect to their absolute values transmitted $1ms$ earlier in message B. Thus message B contains the absolute values of the accelerometers and offsets of the gyroscope signals.

For each offset one byte is reserved, which corresponds to $18.75°/s$ and $250mg$. Thus, for a $1kHz$ frame rate, gradients of up to $18750°/s^2$ and $250g/s$ can be resolved, which is enough for the mechanical systems μCube will be connected to.

The second external interface of μCube uses the RS232 standard. This serial protocol allows for direct connection to PCs, FPGAs or (low-cost) microcontrollers. Special attention has to be paid to the transmission rate. If the IMU needs to support CAN bus, it must be equipped with a $16MHz$ quartz. This quartz, however, does not support all rates accepted by standard PCs [9]. In this setup, baud rates of 250, 76.8, 38.4, 19.2, 9.6, 4.8 and 2.4 $kbit/s$ are supported. For a version without CAN interface, no restrictions apply since the μCube can be populated with a quartz frequency corresponding to RS232 rates.

3.6 Layout and Topology of μCube

3.6.1 System Layout

The highest priority for the system layout had to be a small design envelope with dimensions in any direction lower than $20mm$. This is especially crucial since size is the unique selling point of μCube. To fulfill the requirement of 6dof sensing, the three sensitive axes of both acceleration and angular rates need to be mounted orthogonal to each other. Proper orthogonality not only decreases the necessary filtering burden but also increases sensitivity in any axis in space and thus improves sensing resolution. Nevertheless, mounting has a finite accuracy especially at the intended low production costs. Therefore, compensation is required as proposed in Chapter 5.

In the μCube design, accelerometers do not require special attention since all axes are integrated in one single chip. The used ADXRS series gyroscopes, however, are sensitive in yaw direction and thus require mounting on three orthogonal planes. Most standard commercially available IMUs consist of a main circuit board with two small gyroscope boards orthogonal to it.

μCube goes one step further. To make most efficient use of the 3D design envelope defined by the already three PCBs, it enhances the cube to be six sided and completely enclosed. On these additional three PCBs, all other electronic components that are usually on the "main" board are placed, resulting in the loss of the relatively large board and reduction in the maximum extension.

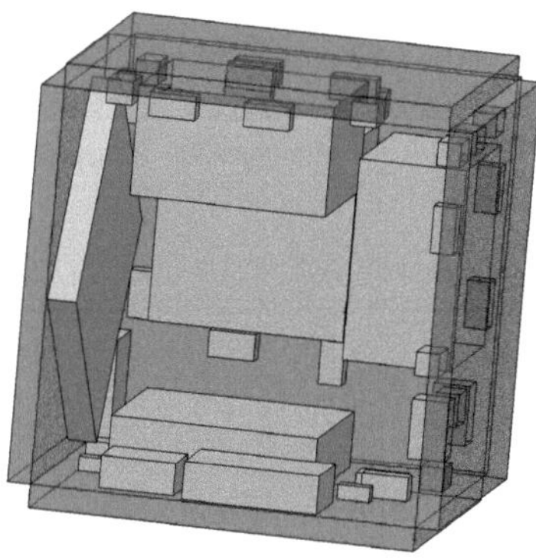

Fig. 3.7: 3D CATIA model for the collision analysis of the six sided cube (one side opened for clarity).

In order to best protect the sensitive electronic components and to ensure better handling of the cube, nearly all components are placed on the inside surface of the boards. Only the interface board has parts on the outer side. This has the additional advantage that energy consuming parts, which usually heat up a bit, can be placed outside the cube. Fig. 3.7 shows a collision analysis model, ensuring efficient use of space inside the cube. The gyroscopes with a height of $5mm$ are especially difficult to place. The final design has an outer edge length of $15mm$, which leads to an $13.6mm$ interior cubical space with the use of $0.8mm$ thick PCBs.

Fig. 3.8 compares the size of μCube to already available IMU/IMS. With its final dimensions of $15mm \times 15mm \times 16mm$ (due to some electronic components on one outer surface of the cube), it saves more than 85% space in comparison to the smallest commercially available system by Cloudcap Technologies. There is only one IMS developed at Tokyo University in between, which is not for sale.

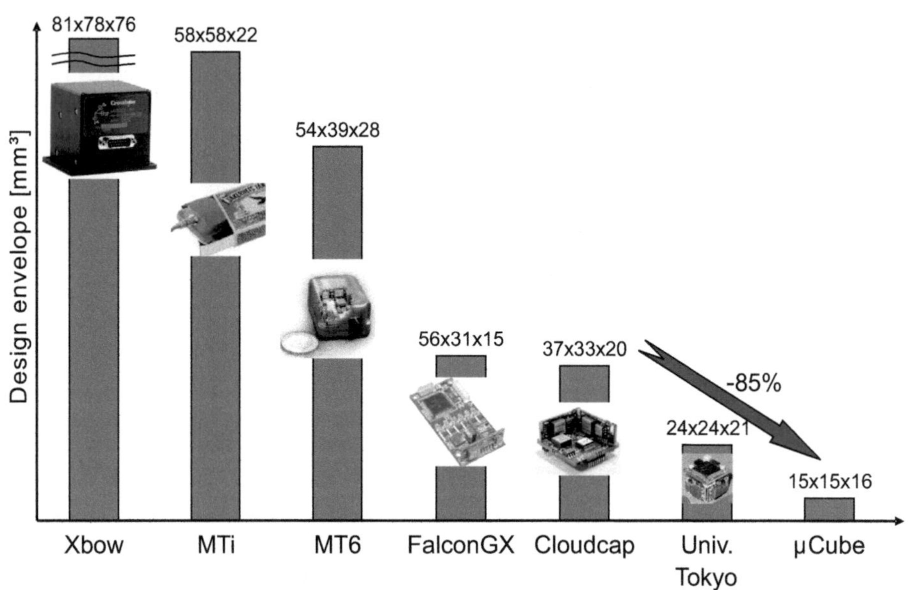

Fig. 3.8: Size comparison of μCube to its main competitors.

With six circuit boards, the electronic interconnections become an important issue. Usually this task is solved by either small connectors or by pairs of soldering pads which go to the edge of each board and then are connected by a flow of tin-solder. Both approaches require too much space for the demands of μCube and therefore different solutions are chosen.

3.6.2 Wirelaid PCB

One approach is the use of so-called wirelaid PCB [143]. In a special manufacturing process small wires with a diameter of $150\mu m$ are placed on the inner surface side of a top level layer for a printed circuit board and laser welded to pads for electrical connection. Then, the PCB is manufactured in the usual standard process. In the final step, small notches are milled out of the PCB and the board is bent [80] [81] around the layer with the subjacent wires, Fig. 3.9. For stabilization, adhesive material is filled in the notches.

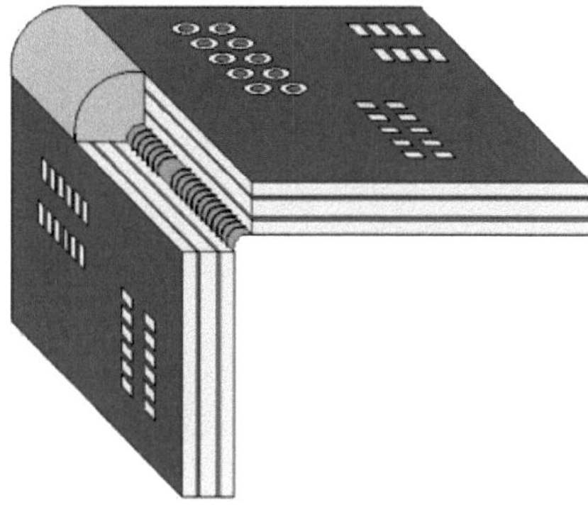

Fig. 3.9: Schematic view of Wirelaid PCB.

In that way, the six sides of the cube are actually manufactured as a single board, which is more advantageous than soldering individual boards together or using connectors:

- no space is required for connectors or soldering pads

- reduced signal routing complexity leads to a more space efficient design

- cost efficient population of one single board

- reduced manufacturing complexity

- electrical connections are robust to vibrations

- electrical connections are protected from environmental influences

Cost reduction in the manufacturing process of the system face higher production costs of the printed circuit board itself. Since the cost of the board reduces significantly with the batch size of the production, whereas savings in the population remain constant, wirelaid PCB is especially suitable for large number production. For this inertial measurement unit technology, a German patent has been filed [59].

Fig. 3.10 shows a half-folded and unpopulated version of μCube.

Fig. 3.10: μCube in wirelaid PCB design.

3.6.3 Soldered Via Connections

For production in lower quantities, a solution for inter-board connections with more manual assembly work and lower fixed costs has been developed. In order not to enlarge the overall dimensions of the cube, a space efficient design within the boards themselves is ideal.

At the edges of each $15mm \times 15mm$ board a high precision 45° chamfer is milled to the board, leaving a $13.4mm \times 13.4mm$ inner square surface. Within the chamfer, electrical vias (Vertical Interconnect Access) are also cut at a 45° angle. When two boards are properly pasted together, these chamfered vias form ducts with a diameter of $0.8mm$. Tin-solder is then inserted into these ducts. Due to the surface treatment of the boards, the solder flows within the ducts and inter-via-connections are avoided.

Fig. 3.11 shows the completely closed μCube with soldered via connections. On the upper and the two lower left edges, several connections are distinguishable. For further stabilization of the cube, $1.2mm$ vias at the corners are soldered together, which makes further adhesive unnecessary.

Fig. 3.11: μCube with soldered via connections.

3.7 Sensor Signal Improvement

Given the final design of the inertial measurement unit, the user has two main possibilities of adapting the sensor signal properties to specific needs. Both of them are based on oversampling.

The Nyquist theorem states that a sensor signal must be sampled at least with the double of the frequencies of interest in the input signal, otherwise the high-frequency content will fold, or alias, at a frequency inside the spectrum of interest:

$$f_{Nyquist} \geq 2 \cdot f_{Signal} \tag{3.1}$$

where f_{Signal} is the highest frequency of interest in the sensor signal. Sampling at higher frequencies than required is referred to as oversampling, defined by the oversampling ratio (OSR)

$$OSR = \frac{f_s}{2 \cdot f_{Signal}} \tag{3.2}$$

where f_s is the sampling frequency. Additional information gained by oversampling can then be used in two ways, to increase the sampling resolution and to decrease the noise level, e.g. increase the signal-to-noise ratio (SNR) [119] if three main criteria are fulfilled [23]:

- The noise must be approximately white noise in the frequency band of interest.

- The noise amplitude must be sufficient to toggle at least one bit of the ADC (1 LSB).

- The sensor signal must have equal probability of existing at any value between two adjacent ADC codes.

Resolution enhancement
With a cycle time of $1ms$ at a $1kHz$ frame rate and only $66\mu s$ needed for the data transfer from the ADC to the microcontroller (Fig. 3.5), the conversion time of a single sample of $4.1\mu s$ allows for an OSR of more than 75 when three gyroscopes are sampled. With a required OSR of 4 for each additional bit of resolution R_{add} [8],

$$R_{add} = \log_4(OSR), \tag{3.3}$$

conversion accuracy for the gyroscopes can be increased by $3bit$ for the μCube. For the acceleration measurement, the LIS3LV02DQ limits the data output rate to $2560Hz$ due to its onboard conversion, resulting in a lower resolution increase, depending on the required system output frame rate.

Noise reduction
If an increase in sensor resolution is not required, oversampling can be used to significantly reduce the output noise level. Again, this works especially well for the angular rate sensors. The used MAX1226 A/D converter provides up to 32x internal oversampling and averaging, allowing for signal improvement without CPU interference. The resulting noise level reduction has been quantified for all available OSRs.

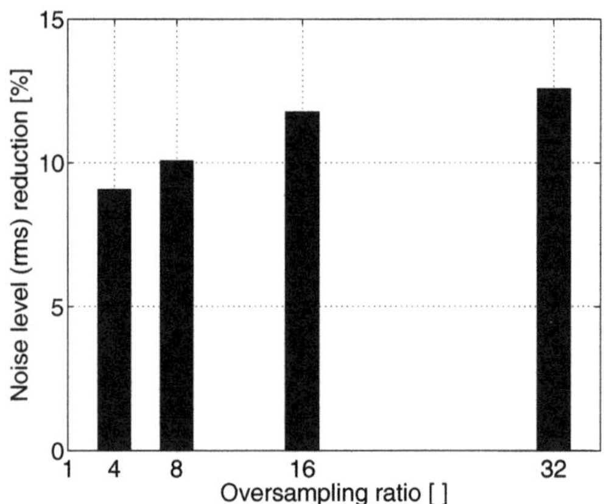

Fig. 3.12: Percental decrease in noise level (rms) for varying oversampling ratios.

Already for 4x oversampling, the noise level could be reduced by more than 9%, increasing up to 13% for an OSR of 32. The insignificant reduction between the last ratios can foster the idea of combining the two methods of signal enhancement, using 8x ADC internal oversampling and averaging for noise reduction and the remaining OSR 4 for increasing the signal resolution by $1bit$.

For the remaining part of this book, an improved gyroscope resolution is not required, however low noise is highly desired. Therefore the gyroscopes will be 32 times oversampled. Due to the already low noise of the accelerometer and the insignificant possible ameliorations at high frame rates, acceleration signals remain untreated.

3.8 Experimental Characterization

This section will focus on the experimental characterization of the μCube. All data is based on tests performed with the cubes with manufacturing series number (MSN) 1 and 2. Emphasis is put on noise and dynamical characteristics especially since sensitivity is subject to high manufacturing tolerances and therefore the characterization of only two sensors would not be significant. In addition, the online adaptation of sensitivity without prior identification is one of the FORBIAS project goals, referred to in Chapter 5.

3.8.1 Acceleration sensors

The LIS3LV02DQ accelerometer that is built in the μCube, combines tri-axial sensing with an onboard ADC and a small control unit. Therefore, only little external influence on the sensor is possible. For the quality of acceleration sensing, the most relevant characteristic is the noise level. Sophisticated adaptive control algorithms, as described in the remaining chapters, can cope with imperfect sensitivity or bias, but noise is a random variable that cannot be predicted.

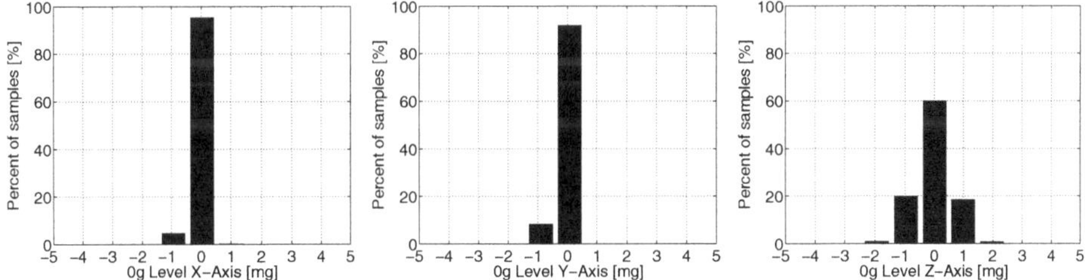

Fig. 3.13: Noise distribution of the LIS3LV02DQ, in sensor plane axes (*left* and *middle*) and out of sensor plane axis (*right*).

Fig. 3.13 visualizes the noise distribution of all three sensitive axes when, within mounting tolerances, subject to no acceleration, e.g. the characterized axis lies stable in a horizontal position.

At a bandwidth setting of $40Hz$, which is the most appropriate for the intended application scenarios, the two in sensor chip plane axes feature a noise level of $0.1mg\,rms$. In more than 90% of all recorded samples, the output is constant at the predominant zero acceleration value. In the remaining samples, the sensor output varies 1LSB, corresponding to approximately 1mg. Noise toggles only one LSB.

Due to the physical design, the out of plane axis of the sensor die has a higher noise level of $0.8mg\ rms$. Zero acceleration output is spread over five bits, leading to a more Gaussian distribution.

The bandwidth of the accelerometer can be dynamically set up to $640Hz$, with is by far enough for all applications of the μCube. For this reason dynamic characterization is limited to the angular rate sensors, which are specified by the manufacturer at only $40Hz$ but shall be used at higher bandwidth settings.

3.8.2 Angular rate sensors

Noise characteristics

In contrast to the accelerometer, the bandwidth of the angular rate sensors has to be set physically with an RC lowpass filter in hardware on the circuit board and cannot be changed after the assembly of the cube. Bandwidth setting has a significant influence on the sensor noise level. At a setting of $40Hz$, which is the reference in the Analog Devices' data sheet, an rms value of only 0.56°/s can be attained, which is lower than specified in the gyroscope data sheet. For higher bandwidth settings, however, the noise level rises almost linearly, Fig. 3.14. Therefore bandwidth should only be set as high as necessary for a specific application.

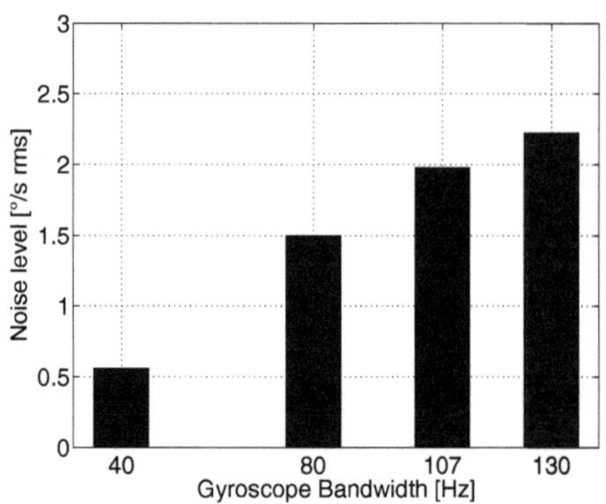

Fig. 3.14: Gyroscope noise level at varying bandwidth settings.

Similar to the accelerometer characterization, Fig. 3.15 visualizes noise distribution for all three gyroscope axes of μCube MSN1. The resolution of the analog-digital converter corresponding to 0.24°/s is precise enough to resolve the noise in a Gaussian distribution.

The low rms noise level of the complete angular rate sensing system, including gyroscopes, analog filtering, and analog-digital conversion is especially remarkable since analog information is passing close to digital signals on the circuit board, given the inner cubical size of $13.4mm$.

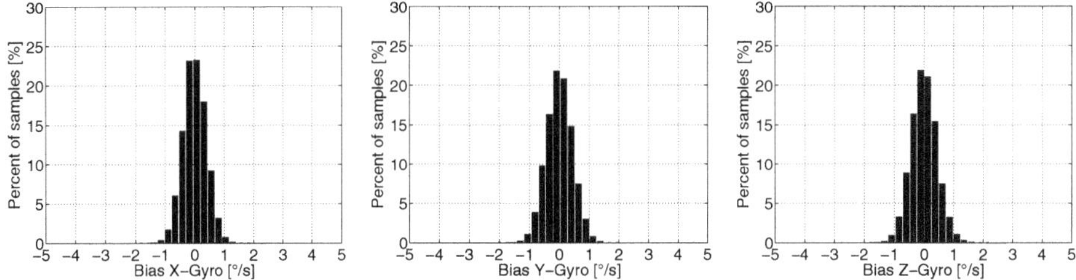

Fig. 3.15: Noise distribution of the three gyroscopes ADXRS300.

Warm-up behavior

In contrast to the accelerometer output, which is comparatively independent of temperature changes, variations from the gyroscopes' specified operating point have significant influence. But since each gyroscope die features an onboard temperature sensor, changes can be tracked and compensated for. Most significant effects occur during the first minutes of operation. Within the first 5 minutes after the gyroscopes

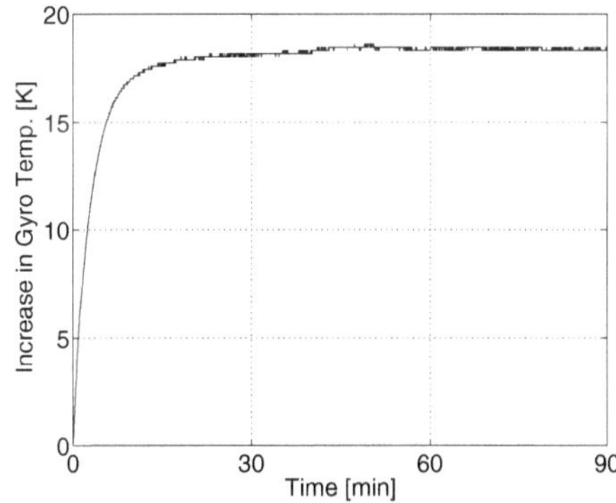

Fig. 3.16: Gyro temperature behavior during warm-up.

are powered, the die temperature rises by more than $15K$, Fig. 3.16. At an ambient temperature of 21 °C, no further change is recorded after the variation peaks at about $18K$ 30min from startup. At that moment, the whole μCube is in thermal equilibrium. The outer surfaces of circuit boards experience an increase of $11K$ from ambient temperature. This is especially relevant since it proves that the electronic components, which are inside the completely closed cube, do not leave the range of their specified operating conditions.

From these changes arise two consequences: variation in sensor sensitivity and bias. Sensitivity alterations are dealt through adaptive algorithms as presented in Chapter 5 and are neglected for spatial orientation determination since multi-sensor fusion algorithms provide sufficient compensation capabilities with others sensors. More relevant for structural compensation is the change in bias. With a given warm-up curve, the bias of a specific sensor consists of a constant value, a time-dependent part and random walk. The latter is stochastical and cannot be predicted but has to be taken into account with probabilistic approaches like in Chapter 4. The first two parts can be described by the equation

$$b = C_1 + C_2(1 - e^{-\frac{t}{C_3}}) \tag{3.4}$$

with time t from start-up and three parameters C [98], which have to be experimentally determined. Therefore the gyroscope output, when not subject to motion, has been recorded for an hour after start-up. The first 5 minutes are plotted in Fig. 3.17 (*left*).

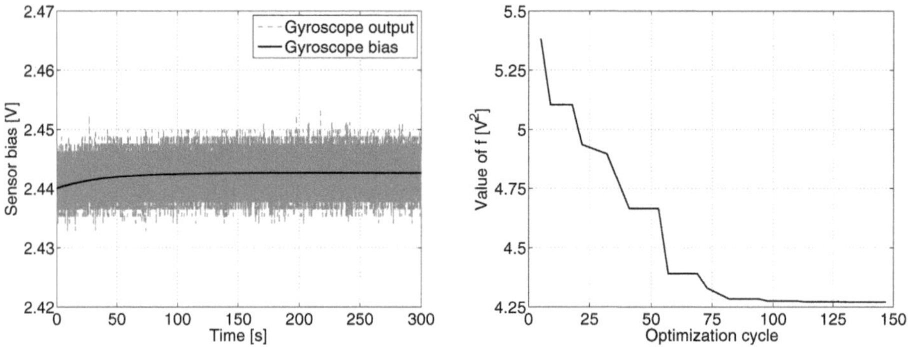

Fig. 3.17: Gyroscope bias after start-up (*left*), optimization history of parameter identification (*right*).

For parameter identification, implicit filtering, a non-linear optimization method with steepest descent algorithm is used. Developed by Kelley [84], implicit filtering for constrained optimization (IFFCO) [85] is especially suitable for noisy problems. The possibility to define search boundaries allows for efficient convergence. As an error function f to be minimized, the quadratic sum

$$f = \sum_{\tau=1}^{\tau_{max}} (V_g - b)^2 = \sum_{\tau=1}^{\tau_{max}} (V_g - C_1 - C_2(1 - e^{-\frac{\tau}{C_3}}))^2 \tag{3.5}$$

with gyroscope voltage output V_g, b as of Eq. 3.4 and a reference period $\tau_{max} = 900000[ms]$ was chosen. With 150 optimization cycles, the quadratic error could be reduced by more than 20% with respect to the first optimized iteration, see Fig. 3.17 (*right*).

Dynamic response

The dynamic behavior of the angular rate sensors is crucial for the performance in many applications. For example, bandwidth was the main reason to switch from MEMS gyroscopes in the design of the orientation sensor of Johnnie [47] to fiber-optical angular rate sensors for the successive humanoid robot LOLA. However in the meantime, MEMS technology has progressed, allowing for bandwidths (-3dB magnitude) above $100Hz$.

Since the gyroscopes' manufacturer Analog Devices Inc. specifies the sensor characteristics only for a corner frequency of $40Hz$ with vague information for higher values [4], a test setup was used to determine the dynamic behavior. In that way, not only the gyroscopes themselves but the electronic system of sensors, analog-digital conversion, microcontroller and CAN bus can be characterized. An electro-dynamic shaker's (1) linear motion, as in Fig. 3.18, is converted to rotation of a beam (2), to which the tested gyroscopes are clamped. In position (A), pure rotation is sensed at the beam-fixed axis, in position (B), the influence of linear acceleration on the gyroscopes can be studied. Optical displacement sensor (3) readings are used to calculate the beam's angular rate either taking into account the nonlinear kinematics or linearized around the horizontal mean position. The setup allows for testing frequencies up to $90Hz$.

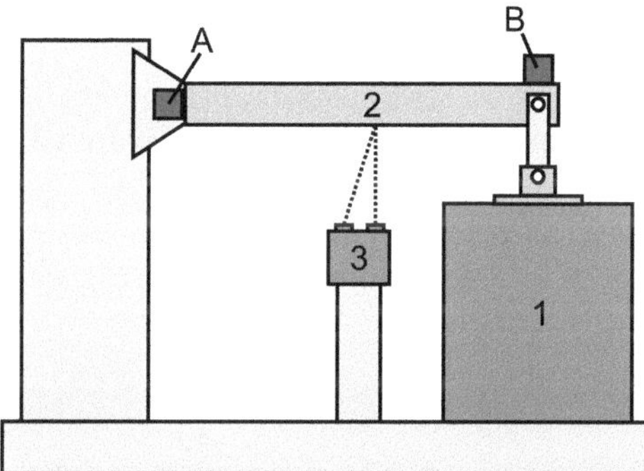

Fig. 3.18: Test set-up for dynamic characterization of μCube's gyroscopes with electro-dynamic shaker (1), rotating beam (2), distance sensor (3) and gyroscope mounting positions (A) and (B).

For the determination of the dynamic behavior of the angular rate sensing system in the form of bode plots, the direct approach is to use gyroscope output and the reference signal from laser triangulation, determine maxima of sinusoidal motion and compare magnitude and phase. For low frequency motion, however, the derivative of the noisy distance signal leads to inaccurate results [118].

Therefore, an analysis in the frequency domain is preferable. Test cycles of $20s$ with sinusoidal motion throughout the measurement range were performed. At each testing frequency f_N, a discrete Fourier transform (DFT) of both gyroscope and distance sensor signal was computed with a fast Fourier transform (FFT) algorithm. The MATLAB algorithm based on FFTW code [42] returns a complex vector with elements corresponding to the frequencies resolved at the used sample time. Information is usable up to the Nyquist frequency, which is half the sampling rate. To eliminate noise, at each test cycle, only the element corresponding to the desired testing frequency f_N is evaluated, for the determination of gain and phase at point f_N in the bode plot of the gyroscope's transfer function.

The magnitude, or gain, at each frequency f_N is determined by the ratio of the absolute values of the corresponding complex entries X_{f_N} and Y_{f_N} in the FFT results vectors of the gyroscope and distance sensor, respectively:

$$G_{f_N} = \frac{\sqrt{\mathrm{Re}^2(X_{f_N}) + \mathrm{Im}^2(X_{f_N})}}{\sqrt{\mathrm{Re}^2(Y_{f_N}) + \mathrm{Im}^2(Y_{f_N})}} \tag{3.6}$$

Similarly, the phase shift in the bode plot, which is the angle difference between the two signals, is calculated at each frequency f_N:

$$Phase_{f_N} = \arctan\left(\frac{\mathrm{Im}(X_{f_N})}{\mathrm{Re}(X_{f_N})}\right) - \arctan\left(\frac{\mathrm{Im}(Y_{f_N})}{\mathrm{Re}(Y_{f_N})}\right) \tag{3.7}$$

Due to the bandwidth setting with a capacitor, it is preferable to choose frequencies that can be attained with one single capacitor in order to not accumulate tolerances of several components, leading to a wide spread in the resulting bandwidth. The low bandwidth range is characterized with two test series with $22nF$ and $11nF$ capacitors, resulting in corner frequencies (-3dB) of $40Hz$ and $80Hz$, respectively.

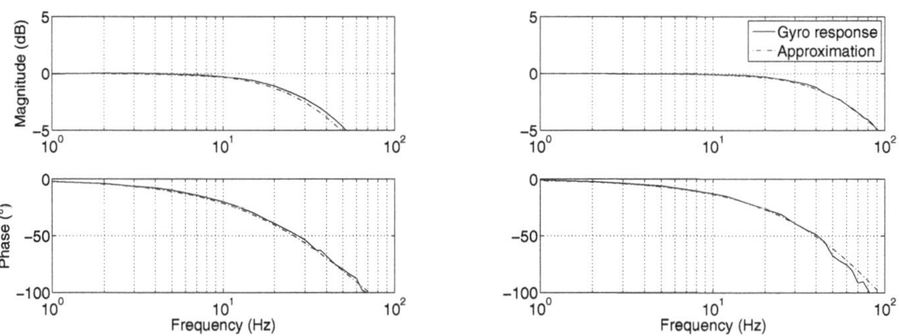

Fig. 3.19: Bode plots for nominal gyroscope settings of $40Hz$ (*left*) and $80Hz$ (*right*), including a PT_2 approximation with delay.

The corresponding bode plots are shown in Fig. 3.19 for frequencies up to $90Hz$, which is the limit imposed by the test setup. For comparison and evaluation, a nominal approximation of the dynamic behavior is plotted with the measurement data. Analog Devices specifies the gyroscopes themselves with a PT_2 characteristic. One internal single-pole (T_I) low-pass filter at $400Hz \pm 35\%$ is used to limit high frequency artifacts before the final amplification stage on the integrated circuit [4]. Then, a second dominant pole (T_B) is set in the application, leading to the effective nominal bandwidth. The transfer function of the gyroscope is determined by

$$G_{gyro} = \frac{S_N}{T_B T_I s^2 + (T_B + T_I)s + 1} \tag{3.8}$$

with the time constants corresponding to the two poles and nominal sensor sensitivity S_N. Due to the high sampling frequency and the high precision of the analog-digital converter, further variations in magnitude do not need to be considered. However, the delays induced by the microcontroller's frame cycles and communication via CAN bus to the master DSP result in a further phase shift. For the sensor system modeling, a first-order Padé approximation of the delay T_Δ [32] is appropriate

$$G_\Delta = \frac{-\frac{T_\Delta}{2}s + 1}{\frac{T_\Delta}{2}s + 1} \tag{3.9}$$

The approximations plotted in Fig. 3.19 as well as in Fig. 3.20 are thus based on the complete angular rate sensing system

$$G_{gyro\_system} = G_\Delta \cdot G_{gyro} = \frac{-\frac{T_\Delta}{2}s + 1}{\frac{T_\Delta}{2}s + 1} \cdot \frac{S_N}{T_B T_I s^2 + (T_B + T_I)s + 1} \tag{3.10}$$

with the individually specified bandwidth setting pole T_B. For all tested bandwidth

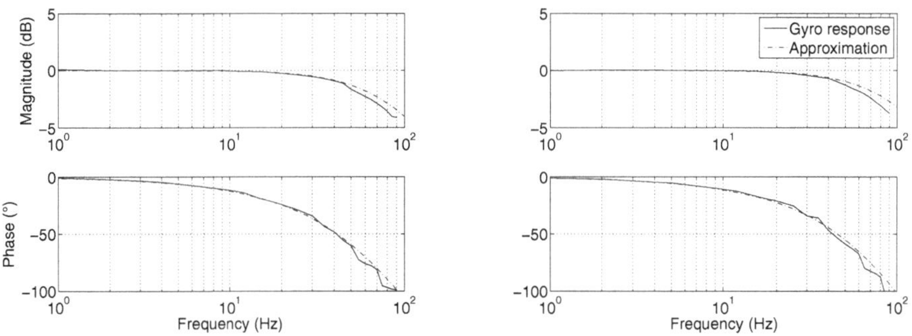

Fig. 3.20: Bode plots for nominal gyroscope settings of $107Hz$ (*left*) and $130Hz$ (*right*), including a PT_2 approximation.

settings up to $130Hz$, the angular rate sensing system met the sensor specification and confirmed the applied model in the tested frequency range up to $90Hz$. The phase behavior is well predicted over the whole range. The loss in gain, especially for higher frequencies in the $130Hz$ setting (Fig. 3.20, *right*) is slightly underestimated. This can be compensated with a slight adaptation of the inverse model in a control application. Taking this into account and noting the predictable phase behavior, the angular rate sensing system of the μCube can be specified and approved for bandwidth settings up to at least $110Hz$.

4 Multi-Sensor Fusion Algorithms for Spatial Orientation Calculation

4.1 Introduction

The determination of spatial orientation is becoming increasingly important in a great variety of applications. In consumer electronics, this is mostly reduced to the resolution of accelerations in a body frame to calculate a device's inclination with respect to the vector of gravity. Current applications include handheld input controllers for maneuvering through virtual reality environments or simple gaming with Nintendo's *Wii Console*. Sony's *PS3*, which is to become commercially available in the course of 2007, apparently uses gyroscopes in addition to accelerometers.

In more serious applications, determination of spatial orientation is a key requirement for navigation. Up until now, navigation systems in automobiles have relied on GPS information coupled with knowledge on the vehicle's velocity and yaw rate, which is determined by steering angles and/or yaw gyroscopes. For future applications, especially in connection with high-level terrain maps, which are considered to play an important role in competitive driver assistance systems [11] [93], complete determination of position and orientation are required. Fiat Research Center published their intention to develop a 6DOF inertial measurement unit with a design envelope of $20mm$ in each direction [27]. In research driven automotive applications, like the DARPA Grand Challenge, autonomously driving cars rely on this information [130].

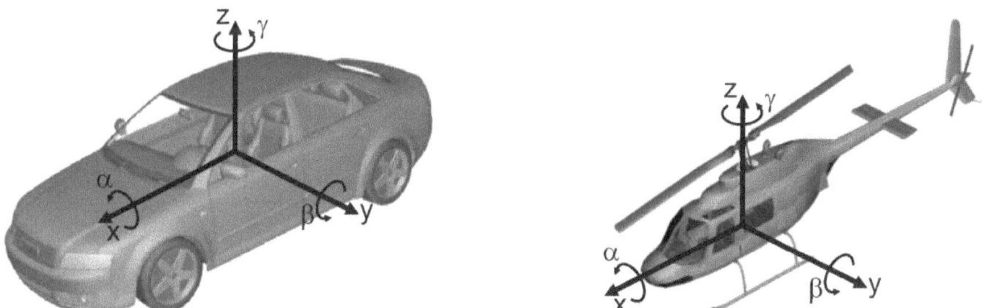

Fig. 4.1: Applications with need of spatial orientation. Vehicle motion is referred to as roll (x), pitch (y) and yaw (z) (non-ISO for aerospace applications).

W. Günthner: *Enhancing Cognitive Assistance Systems with Inertial Measurement Units*, Studies in Computational Intelligence (SCI) **105**, 49–77 (2008)
www.springerlink.com

In aeronautics, the necessity for accurate position and orientation determination is self-explanatory. When referring to these topics, the frame of reference in which navigation is done needs to be specified.

The *inertial frame* (I-frame) has its origin at the center of the Earth with two axes on the equatorial plane. It is non-rotating with respect to fixed stars.
The *Earth frame* (E-frame) has the same properties, except for the fact that it is rotating with the Earth. The x_E-axis lies along the intersection of the Earth's equatorial plane with the plane of the Greenwich meridian, z_E-axis coincides the Earth's polar axis.
The *navigation frame* (N-frame) is a local geographic frame which has its origin usually on the Earth's surface, with its axes pointing North, East, and towards the center of the Earth.
The *body frame* (B-frame) is finally a reference frame attached to a moving body or vehicle.

Before going into details of the different algorithms for orientation calculation by multi-sensor fusion, a short introduction to basic navigation equations for position determination can point out the strategy to enhance position based assistance systems with inertial data. The basic idea is to use GPS positioning as a long-term reference. However, to reduce error and address satellite outages, artificially induced in GPS navigation by the satellite operator, inertial data (e.g. the double integration of vehicle acceleration) can significantly contribute to navigation accuracy.

The accelerometers of an on-board inertial measurement unit resolve the specific forces $\tilde{\mathbf{a}}$ acting on the vehicle

$$\tilde{\mathbf{a}} = \left.\frac{d^2\mathbf{r}}{dt^2}\right|_I - \mathbf{g}, \tag{4.1}$$

but due to the acceleration characteristic of gravity $\mathbf{g}$, only the difference of actual acceleration due to motion in the inertial reference frame I, $\left.\frac{d^2\mathbf{r}}{dt^2}\right|_I$, and gravity can be determined (note that until Eq. 4.8 coordinate-free representation is used). $\mathbf{r}$ denotes the vector from the center of the Earth to the vehicle. In order to arrive at variables that can be measured by the instruments of a vehicle, this vector is differentiated in the E-frame, requiring the theorem of Coriolis:

$$\left.\frac{d\mathbf{r}}{dt}\right|_I = \left.\frac{d\mathbf{r}}{dt}\right|_E + \boldsymbol{\omega}_{IE} \times \mathbf{r} \tag{4.2}$$

where $\left.\frac{d\mathbf{r}}{dt}\right|_E$ is the *local* derivative of $\mathbf{r}$ and $\boldsymbol{\omega}_{IE}$ the turn rate of the Earth frame with respect to the I-frame. When substituting the velocity in E-frame

$$\mathbf{v}_E = \left.\frac{d\mathbf{r}}{dt}\right|_E, \tag{4.3}$$

which could be measured at the wheels of a car, and differentiating a second time, the (total) acceleration in I-frame can be expressed as

$$\frac{d^2\mathbf{r}}{dt^2}\bigg|_I = \frac{d\mathbf{v}_E}{dt}\bigg|_I + \frac{d}{dt}(\boldsymbol{\omega}_{IE} \times \mathbf{r})\bigg|_I \tag{4.4}$$

Applying the theorem of Coriolis to the second term, Eq. 4.4 yields

$$\frac{d^2\mathbf{r}}{dt^2}\bigg|_I = \frac{d\mathbf{v}_E}{dt}\bigg|_I + (\boldsymbol{\omega}_{IE} \times \mathbf{v}_E) + \boldsymbol{\omega}_{IE} \times (\boldsymbol{\omega}_{IE} \times \mathbf{r}), \tag{4.5}$$

assuming a constant turn rate of the Earth, $\boldsymbol{\omega}_{IE} = const$. When substituting with acceleration of Eq. 4.1, the accelerations sensed by the IMU are introduced:

$$\frac{d\mathbf{v}_E}{dt}\bigg|_I = \tilde{\mathbf{a}} + \mathbf{g} - (\boldsymbol{\omega}_{IE} \times \mathbf{v}_E) - \boldsymbol{\omega}_{IE} \times (\boldsymbol{\omega}_{IE} \times \mathbf{r}) \tag{4.6}$$

For navigation represented in GPS world coordinates, this equation would be fine. When performing simultaneous localization and mapping (SLAM), however, a representation in the local navigation frame is often desirable for the extraction of 2D maps. Therefore, the vehicle velocity $\mathbf{v}_E$ is differentiated with respect to the navigation frame N:

$$\frac{d\mathbf{v}_E}{dt}\bigg|_N = \frac{d\mathbf{v}_E}{dt}\bigg|_I - (\boldsymbol{\omega}_{IN} \times \mathbf{v}_E) \tag{4.7}$$

with $\boldsymbol{\omega}_{IN} = \boldsymbol{\omega}_{IE} + \boldsymbol{\omega}_{EN}$. Note that $\boldsymbol{\omega}_{IE}$ is known and $\boldsymbol{\omega}_{EN}$, the turn rate of the navigation frame with respect to the Earth can be derived from the variation of the position on Earth. Since land vehicles move relatively slowly, the last term can often be neglected. When substituting with Eq. 4.6, the acceleration with respect to the navigation frame can be expressed by

$$\frac{d\mathbf{v}_E}{dt}\bigg|_N = \tilde{\mathbf{a}} - (2\boldsymbol{\omega}_{IE} + \boldsymbol{\omega}_{EN}) \times \mathbf{v}_E + \mathbf{g} - \boldsymbol{\omega}_{IE} \times (\boldsymbol{\omega}_{IE} \times \mathbf{r}) \tag{4.8}$$

All variables are known or measurable. If the last two terms are grouped to what is called the local gravity vector $\mathbf{g}_l$ and the equation is transformed from coordinate-free representation to a representation in the N-frame, Eq. 4.8 yields

$$_N\dot{\mathbf{v}}_E = \mathbf{C}_{NB}\,_B\tilde{\mathbf{a}} - (2\,_N\boldsymbol{\omega}_{IE} +_N \boldsymbol{\omega}_{EN}) \times_N \mathbf{v}_E +_N \mathbf{g}_l \tag{4.9}$$

It is obvious that all components except the specific forces, as measured by the accelerometer, are known in the N-frame. The accelerometer output must be transformed from B to the N-frame with the corresponding direction cosine matrix $\mathbf{C}_{NB}$.

This theoretical approach shows the necessity to determine the orientation of a body for the integration of measured specific forces into self localization and positioning. The structure of positioning systems in visualized in Fig. 4.2.

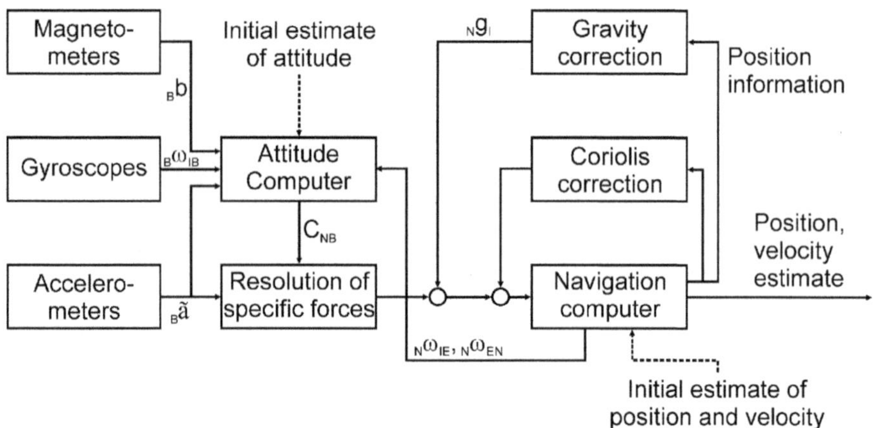

Fig. 4.2: Local navigation frame mechanisation.

The output, position and velocity estimates, are of high quality in the short term range. Due to the double integration of accelerations, they are, however, subject to drift. Therefore, the result of inertial navigation is ideally suited for fusion with GPS data, for example in an Extended Kalman Filter.

After this general introduction to inertial navigation, the following sections will focus on the actual derivation of vehicles' orientation in space. Section 4.2 summarizes an algorithm for restricted motion without (long-term) calculation of heading. Section 4.3 then derives probabilistic approaches to multi-sensor fusion for the calculation of the complete orientation of a vehicle.

4.2 Complementary Filtering for Attitude Reference Systems

Complementary Filtering is a simple method of multi-sensor fusion, optimized for restricted motion and can be derived from a standard Kalman Filter [10] [98]. The necessary linearization is possible if low amplitudes for roll and pitch motion (Fig. 4.1) can be assumed since every coordinate is then influenced by only one sensor bias [47]. The algorithm has low computational demands and can be implemented on almost any microcontroller. For clarity reasons, and since it is usually not required in low-cost applications, heading calculation is exempt from this section. If required, it could be easily added by using magnetometer output coupled with the heading calculation procedure as described in Section 4.3 as a long-term reference, similar to accelerometer output in this section.

With small amplitudes in motion with respect to the navigation frame, this attitude algorithm is well suited for use with land vehicles, especially cars driving on standard roads.

4.2.1 Attitude Representation

In order to clearly identify the orientation of a body, a distinct attitude representation must be defined. In automotive applications, mechanization with respect to a navigation frame, which has its x- and y-axis in the horizontal plane, and the z-axis pointing upwards in vertical direction, is appropriate. Usually, the x_N-axis is oriented in the northern direction, Fig. 4.3.

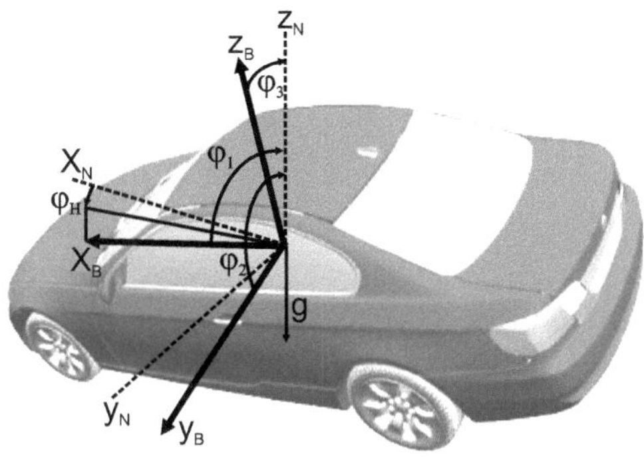

Fig. 4.3: Navigation frame mechanisation.

The angles φ_1, φ_2 and φ_3 describe the attitude of the vehicle, with φ_3 being redundant. Heading φ_H is counted between the projection of the x_B-axis onto the $x_N - y_N$ plane and the x_N-axis. The first three angles are the angles between the three axes of the body frame B and the z_N-axis. In case no additional acceleration is acting on the vehicle and the body is only subject to gravity $\mathbf{a} = \mathbf{g}$, attitude could be directly calculated from accelerometer measurement.

$$\begin{pmatrix} \varphi_1 \\ \varphi_2 \\ \varphi_3 \end{pmatrix} = \begin{pmatrix} \arccos\left(\frac{\tilde{a}_x}{|\tilde{\mathbf{a}}|}\right) \\ \arccos\left(\frac{\tilde{a}_y}{|\tilde{\mathbf{a}}|}\right) \\ \arccos\left(\frac{\tilde{a}_z}{|\tilde{\mathbf{a}}|}\right) \end{pmatrix} \tag{4.10}$$

with

$$|\tilde{\mathbf{a}}| = \sqrt{\tilde{a}_x^2 + \tilde{a}_y^2 + \tilde{a}_z^2} \tag{4.11}$$

and $\varphi_1, \varphi_2, \varphi_3 \in [0, \pi]$. For efficient and accurate attitude calculation, the Jacobian matrix $_B\mathbf{J}$ is needed to transform angular rates $_B\tilde{\boldsymbol{\omega}}$ as measured by the body-fixed gyroscopes to variations in these generalized coordinates:

$$\begin{pmatrix} \dot{\varphi}_1 \\ \dot{\varphi}_2 \\ \dot{\varphi}_H \end{pmatrix} = \underbrace{\begin{pmatrix} \frac{d_B\boldsymbol{\omega}}{d\dot{\varphi}_1} & \frac{d_B\boldsymbol{\omega}}{d\dot{\varphi}_2} & \frac{d_B\boldsymbol{\omega}}{d\dot{\varphi}_H} \end{pmatrix}^T}_{_B\mathbf{J}} {}_B\tilde{\boldsymbol{\omega}} = \begin{pmatrix} 0 & \sin\varphi_2 & -\cos\varphi_2 \\ -\sin\varphi_1 & 0 & \cos\varphi_1 \\ 0 & A^* & B^* \end{pmatrix} \begin{pmatrix} \omega_x \\ \omega_y \\ \omega_z \end{pmatrix} \quad (4.12)$$

with

$$A^* = \frac{\operatorname{sgn}(\pi/2 - \varphi_2)\sqrt{-\cos(\varphi_1 - \varphi_3)\cos(\varphi_1 + \varphi_3)}}{\sin^2\varphi_1} \quad (4.13)$$

$$B^* = \frac{\operatorname{sgn}(\pi/2 - \varphi_3)\sqrt{-\cos(\varphi_1 - \varphi_2)\cos(\varphi_1 + \varphi_2)}}{\sin^2\varphi_1} \quad (4.14)$$

As mentioned above, heading φ_H can not be derived from acceleration measurement but requires additional sensors like magnetometers or dual-antenna GPS. If the heading is derived from $\dot{\varphi}_H$

$$\varphi_H = \int \dot{\varphi}_H \, dt, \quad (4.15)$$

it is subject to drift. In applications where only short-term statements about variations in heading are required, for example for yaw-control, the integrator of Eq. 4.15 can be stabilized with a leakage term with loss rate α

$$\varphi_H = \int \dot{\varphi}_H \, dt - \alpha\varphi_H, \quad (4.16)$$

which results basically in a high-pass if Laplace transformed. For the use of this non-standard attitude description with other navigation systems, an interface for vector multiplication is needed. The most general one is a direction cosine matrix (DCM), or transformation matrix, between the body and the navigation frame. It contains the cosines of the angles between all involved axes, Eq. 4.17.

$$A_{NB} = \begin{pmatrix} a_{11} & a_{12} & a_{13} \\ a_{21} & a_{22} & a_{23} \\ a_{31} & a_{32} & a_{33} \end{pmatrix} = \begin{pmatrix} \cos(\mathbf{x}_B, \mathbf{x}_N) & \cos(\mathbf{y}_B, \mathbf{x}_N) & \cos(\mathbf{z}_B, \mathbf{x}_N) \\ \cos(\mathbf{x}_B, \mathbf{y}_N) & \cos(\mathbf{y}_B, \mathbf{y}_N) & \cos(\mathbf{z}_B, \mathbf{y}_N) \\ \cos(\mathbf{x}_B, \mathbf{z}_N) & \cos(\mathbf{y}_B, \mathbf{z}_N) & \cos(\mathbf{z}_B, \mathbf{z}_N) \end{pmatrix} \quad (4.17)$$

The angles between the body axes and the inertial z-axis are known from the definition,

$$a_{31} = \cos\varphi_1, \quad a_{32} = \cos\varphi_2, \quad a_{33} = \cos\varphi_3 \quad (4.18)$$

With heading additionally derived from magnetic field evaluation (section 4.3) or by integration (Eq. 4.16), the first column of the DCM can be completed:

$$a_{11} = \sin\varphi_1 \cdot \cos\varphi_H, \quad a_{21} = \sin\varphi_1 \cdot \sin\varphi_H \quad (4.19)$$

Since all coordinate systems are orthogonal with basis vectors of unity length, the second column is found with

$$a_{11} \cdot a_{12} + a_{21} \cdot a_{22} + a_{31} \cdot a_{32} = 0 \quad and \quad a_{12}^2 + a_{22}^2 + a_{32}^2 = 1 \qquad (4.20)$$

leading to two mathematical solutions of the resulting quadratic equation. The third is the cross product of the first two columns. With the restriction that the body z-axis must not point downwards, which would mean the car is lying upside down, the required positive value of a_{33} eliminates the wrong mathematical solution to the second column of the DCM.

This representation is used throughout Section 4.2. In addition, attitude calculation from accelerations, as described above, will be used in the entire chapter.

4.2.2 Filter Algorithm

The derivation of attitude angles with Eq. 4.10 is valid only in case no other acceleration except gravity is present (e.g. the car is at a complete standstill). When moving, bumps and vibrations temporarily degrade the result such that accuracy expectations cannot be met. On the other hand, attitude angles φ_1 and φ_2 can be propagated accurately with Eq. 4.12 for a short time until angular rate integration drift becomes predominant.

The basic idea of complementary filtering is to use each kind of information in the range where it has the highest accuracy, a straightforward approach to multi-sensor fusion. Gyroscope signals are thus best for short term and high frequency ranges, and the accelerometer reading is used for long term stability to avoid drift in φ_1 and φ_2. Fig. 4.4 outlines the principle of complementary filtering.

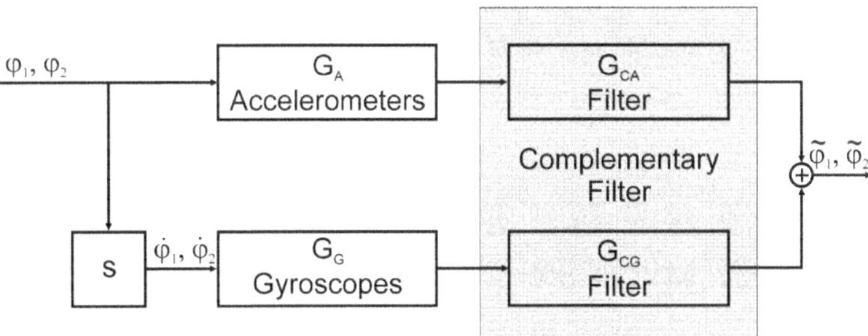

Fig. 4.4: Basic principle of complementary filtering.

Attitude angles, or the corresponding accelerations, are sensed by the accelerometers (Eq. 4.10), characterized by their dynamic transfer function G_A. The change in attitude (e.g. rotation) is detected and propagated (Eq. 4.12) in the gyroscopic path with G_G. In the complementary filter, the acceleration branch is then lowpass filtered with G_{CA}, the gyroscope branch highpass filtered. Additionally, G_{CG} includes integration, since its input corresponds to $\dot{\varphi}_1$ and $\dot{\varphi}_2$ (Eq. 4.12).

The goal is a transfer function between the estimated $\tilde{\varphi}_1$, $\tilde{\varphi}_2$ and actual attitude φ_1, φ_2 of unity, e.g. for φ_1:

$$G_{CF} = \frac{\tilde{\varphi}_1}{\varphi_1} = G_A \cdot G_{CA} + s \cdot G_G \cdot G_{CG} = 1 \tag{4.21}$$

With a bandwidth of more than $150Hz$, the dynamic behavior of accelerometers can be assumed to be ideal in the considered range of vehicle motion (Section 3.1), i.e. $G_A = 1$. If delays can be neglected due to onboard calculation on the IMU's microcontroller, G_G can be derived from Eq. 3.8

$$G_G = \frac{1}{T_B T_I s^2 + (T_B + T_I)s + 1}, \tag{4.22}$$

shifting S_N to the calculation of $\tilde{\varphi}_1$ and $\tilde{\varphi}_2$ and considering just the dynamic behavior of the gyroscopes. With only one equation (4.21) and G_{CA}, G_{CG} unknown, there is one degree of freedom left. Baerveldt [10] proposes a second order filter with double pole for a similar problem, reducing the design parameters to only one time constant T_s

$$G_{CA} = \frac{2T_s s + 1}{(T_s s + 1)^2} \tag{4.23}$$

Substitution with Eq. 4.23 in Eq. 4.21 yields the required complementary transfer function

$$G_{CG} = \frac{1 - G_{CA}}{s \cdot G_G} = \frac{T_s^2 s(T_B T_I s^2 + (T_B + T_I)s + 1)}{(T_s s + 1)^2} \tag{4.24}$$

The shape of the filter's amplitude, or gain is plotted with respect to the input frequency in Fig. 4.5. The only filter tuning parameter is the common time constant T_s, the separation frequency. It must be determined with respect to the expected motion input signals. If it is set too high, bumps and shocks have an important degenerative influence on the result. Set low, the initial settling time of the filter increases, and in the worst case scenario gyroscope drift increases.

To get the order of magnitude of T_s, Fig. 4.6 shows the Fast Fourier Transformation, the frequency spectrum of accelerations as detected by an inertial measurement unit during a test drive on a motorway at varying car velocities. The first eigenfrequency of the test vehicle, an AUDI A8, is visible at about $2Hz$. Higher frequency excitation corresponds to the spinning of the wheels (15-20Hz) and motor vibration (50-60Hz) at the testing car velocity.

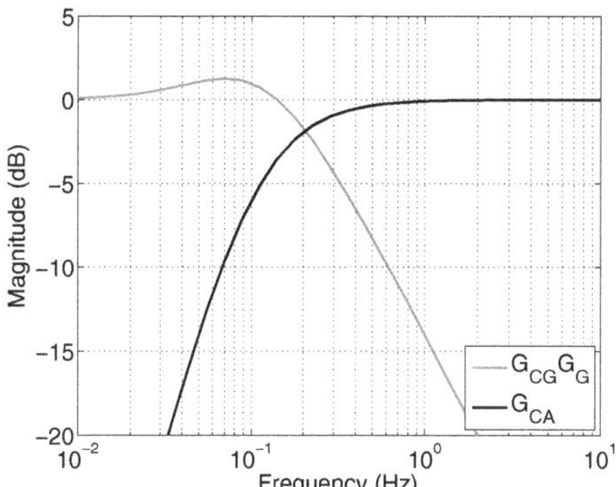

Fig. 4.5: Filtering characteristics of both branches, with angular rate branch corrected by gyroscope compensation G_G, $T_s = 0.1Hz$.

Fig. 4.5 and Fig. 4.6 suggest a time constant T_s corresponding to $0.1Hz$. Accelerations peaking at $2Hz$ suffer from a gain loss of $-20dB$ in the vicinity of the car eigenfrequency, which is sufficient to filter out such influences on orientation determination.

The structure of the filters has basically an influence on the decay of the gain. In this case, the achieved $-20dB$ per decade for filtering accelerations is sufficient.

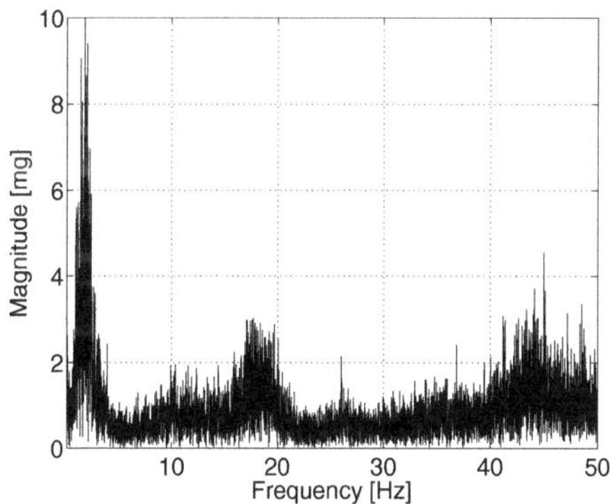

Fig. 4.6: FFT of accelerations at the car frame, $f_{sample} = 1kHz$.

4.2.3 Performance Evaluation

During test drives (Chapter 6.1), the true attitude of the car could not be determined by an external reference system since test drives were performed on public motorways. In order prove the performance of complementary filtering under the influence of realistic disturbances, the following model considers measured sensor values to a high extent.

A coupled sinusoidal motion in φ_1 and φ_2 is the simulation input, see Fig. 4.7, which is used to calculate accelerations $_B\mathbf{a}$ and angular rates $_B\boldsymbol{\omega}$ as detected by body fixed accelerometers and gyroscopes. As a motion amplitude, $\pm 5°$ in pitch corresponds to regular driving. Roll motion is less significant, so it is represented with a $\pm 2°$ amplitude. Typical frequencies in the range of the first eigenfrequencies of rotation were chosen. The sensor output is then mixed with disturbances like noise and vibrations actually recorded in the test vehicle, thus including excitations in the relevant modes.

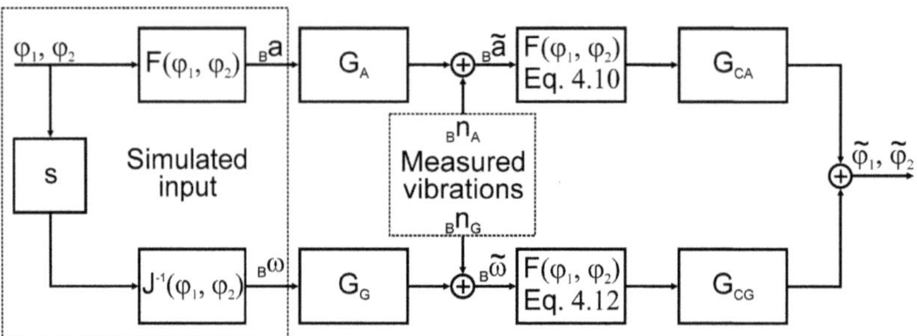

Fig. 4.7: Simulation model of the complementary filter for performance evaluation.

On this half-simulated, half-measured data, the complementary filter is applied. For comparison, simulations show the effects of increasing the common pole time constant to a value corresponding to $0.5Hz$, in addition to the desired $0.1Hz$. Fig. 4.8 (*left*) shows the initial settling behavior of the complementary filter.

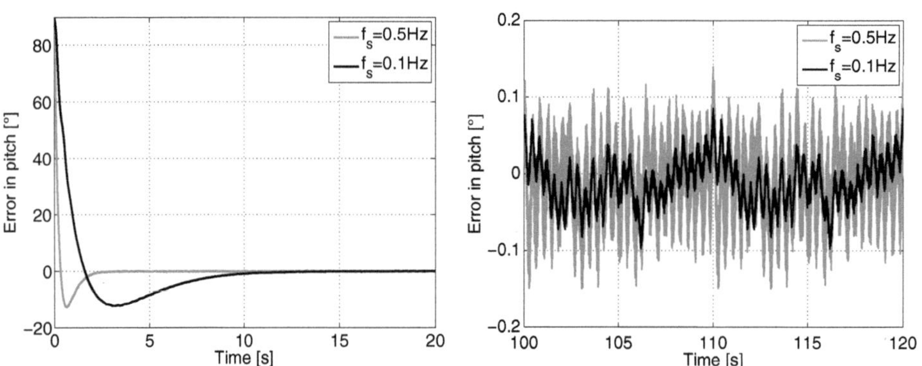

Fig. 4.8: Error in pitch angle calculation during the initial 20s after startup (*left*) and during regular operation (*right*) for different separation frequencies of the complementary filter.

As expected, the filter with the higher time constant requires more time to settle, however with less than $20s$ being in an acceptable range for most applications. In terms of final operation performance, a significant difference can be observed. A separation frequency of $f_s = 0.1 Hz$ filters out more of the vibrations of the car. The higher weight of gyroscopes reduces the rms error in pitch calculation to $0.03°$, down 40% in comparison to $f_s = 0.5 Hz$. Peak error depends on the current coupling of pitch and roll motion, but does not exceed $0.1°$.

4.3 Stochastic Approaches to Orientation Determination

In contrast to the previous section, which focused on restricted motion, this part will enlarge the scope to unrestricted cases. Although the theoretical background will be applied to orientation determination, the presented methods are suitable for a wide range of multi-sensor fusion and integration. The foundation for these approaches is the Bayes framework of conditional probability. It mathematically takes into account prior knowledge or estimates of the state and fuses them with new data, always with respect to the level of confidence in the information. In fact, this is quite similar to the human system and behavior, which also considers prior knowledge in more or less important ways.

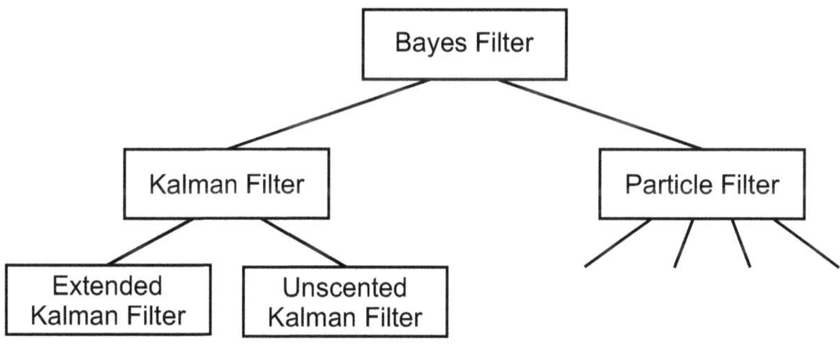

Fig. 4.9: Probabilistic Methods for sensor data fusion [129].

Within the Bayes theory, two main categories have to be distinguished. Kalman Filters can be applied for Gauss-Markov processes, which allow predictions of a next state without having to know the whole history of the process. This is especially suitable for technical implementations since there is no need for storing all prior states in a memory. Until now, the most popular implementation is the Extended Kalman Filter (EKF), which is a nonlinear extension of the standard Kalman Filter. For highly nonlinear processes, Sigma-Point Kalman Filters (SPKF) are currently proposed and in development. Their main advantage is the truly nonlinear propagation of uncertainties, especially noise, with the unscented transformation.

In case process and noise cannot be assumed to be Gaussian, for example if the probability density has more than one maximum, Particle Filter (PF) can be used to propagate states with the use of sample points (particles). Especially in combination with a high number of particles, the algorithm can become computationally expensive. Therefore, it is often just applied to the most nonlinear part of a system and combined with regular EKFs. Apart from these so-called Rao-Blackwellized Particle Filters, a great variety of derivations exist.

What all approaches have in common, is that they are used for processes in which not all states can be observed or measured. In our case, the orientation of a body cannot be measured with sensors, like for example encoders, only some effects, like acceleration or magnetic field are sensed. The applied probabilistic approaches can cope with this handicap. After an extension of orientation representation with heading information, Sections 4.3.2 - 4.3.4 will present three different algorithms for orientation determination that will be compared in the sections after.

4.3.1 Quaternion based navigation frame representation

Up until now, attitude representation with respect to a horizontal frame was useful for low amplitude motion, especially when a heading is not required for calculation. For unrestricted orientation, when a heading is generally of interest, different representation methods are preferable. This is especially the case when orientation data has to be exchanged with external systems which require standard representation like Kardan/Euler angles or quaternions. The latter have several major advantages for inertial navigation. Quaternions allow for singularity-free calculation of all possible orientation states. Kardan and Euler angles would require additional software overhead to detect possible singularities and to switch to a different system if necessary. All this can be avoided by the use of quaternions. In addition, the number of mathematical operations for a rotation sequence is lower than for angular descriptions, which need consecutive multiplication with transformation matrices.

For the calculation of attitude and heading, a navigation frame with horizontal x_N and y_N-axis is defined in a way the x_N-axis is pointing towards magnetic north. In case heading with respect to geographic north is required, the known fixed offset for each point on Earth can be added. Similar to Eq. 4.10, the angles φ_4, φ_5 and φ_6 represent the orientation of the body-fixed coordinate system with respect to the vector of the magnetic field $\mathbf{B}_{GM}$, Fig. 4.10.

$$\begin{pmatrix} \varphi_4 \\ \varphi_5 \\ \varphi_6 \end{pmatrix} = \begin{pmatrix} \arccos\left(\frac{_B B_x}{|\mathbf{B}_{GM}|}\right) \\ \arccos\left(\frac{_B B_y}{|\mathbf{B}_{GM}|}\right) \\ \arccos\left(\frac{_B B_z}{|\mathbf{B}_{GM}|}\right) \end{pmatrix} \tag{4.25}$$

with the intensity of the geomagnetic field

$$|\mathbf{B}_{GM}| = \sqrt{{}_B B_x^2 + {}_B B_y^2 + {}_B B_z^2}$$ (4.26)

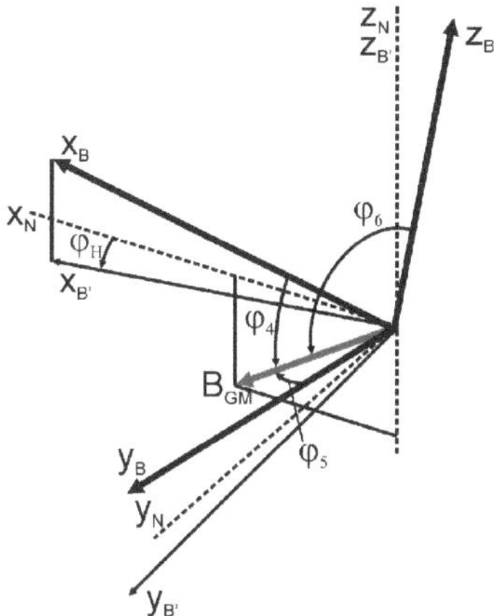

Fig. 4.10: Magnetic navigation frame representation.

These three angles, with φ_6 being redundant, are a second way of representing information on the body's orientation, after calculation from accelerations (Section 4.2.1). Again, measurement is highly sensitive to disturbances, this time from external magnetic fields, and the representation is not complete either, since there is still one degree of freedom, rotation around the axis of the magnetic field vector. However, attitude and heading calculation contain different information than acceleration measurement and is therefore highly suitable for multi-sensor fusion.

But the main reason for magnetic field measurement is determination of heading φ_H. Therefore, an intermediate coordinate system B' with two axes in the horizontal plane, and $x_{B'}$ being the vertical projection of x_B is introduced. It can be derived from B by a two elementary rotation transformation with a rotation of Θ_1 about x_B and Θ_2 about the new $y_{B'}$-axis:

$$\Theta_1 = \arctan\left(-\frac{\cos\varphi_1}{\cos\varphi_3}\right) \quad \text{and} \quad \Theta_2 = \arcsin(\cos\varphi_2)$$ (4.27)

The components of the projection of the magnetic field vector B_{GM} into this horizontal system B' are given by

$${}_{B'}B_{GM,x} = {}_B B_x \cdot \cos\Theta_1 + {}_B B_z \cdot \sin\Theta_1$$ (4.28)

$${}_{B'}B_{GM,y} = {}_B B_x \cdot \sin\Theta_1 \cdot \sin\Theta_2 + {}_B B_y \cdot \cos\Theta_2 - {}_B B_z \cdot \cos\Theta_1 \cdot \sin\Theta_2$$ (4.29)

With these components and Fig. 4.10, heading φ_H can be calculated

$$\varphi_H = \arctan\left(\frac{{}_{B'}B_{GM,y}}{{}_{B'}B_{GM,x}}\right) \tag{4.30}$$

Similar to Section 4.2.1, the direction cosine matrix is calculated from φ_4, φ_5 and φ_H. For further use with the probabilistic algorithms in the next sections, the DCM is transferred to quaternion representation [88].

4.3.2 Extended Kalman Filters for multi-sensor fusion

The popular Extended Kalman Filter is based on a *Predictor-Corrector* principle. A known or estimated state of the system x_{t1}, for example orientation at time t_1, is propagated with modeled system dynamics and input variables to the next time step t_2^-. Then an actual observation/measurement $z(t_2)$ is used for the update of the state, leading to the corrected output $x(t_2^+)$.

Crucial for this type of multi-sensor fusion is the weight each type of information, the prediction (prior) or the measurement, is attributed. In statistic approaches, the variance of a value is a measure for the level of reliability. Fig. 4.11 visualizes the changes in variance for a single scalar state. During propagation with time and possible uncertainties in the system equations, the level of confidence decreases, leading to an increased variance $\sigma_x(t_2^-)$ of the prior at t_2. The measurement itself is characterized by $\sigma_m(t_2)$. When the prior is updated/corrected with the measurement, the posterior state is a weighted average of the former two.

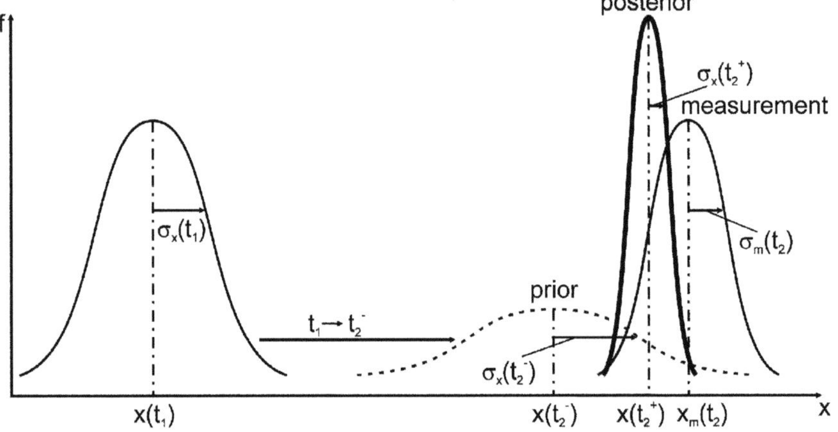

Fig. 4.11: Propagation and update of probability density f for state x.

Note that the variance of the posterior, $\sigma_x(t_2^+)$, is smaller than each of the other two variances at t_2, signifying an increased level of confidence in the posterior. In fact, this is an important observation, since it signifies that the consideration of a second source of information, no matter how unreliable it is, improves the estimation output.

This approach shall now be extended to the orientation determination with several state variables. Fig. 4.12 gives an overview of the multi-sensor fusion process with an Extended Kalman Filter.

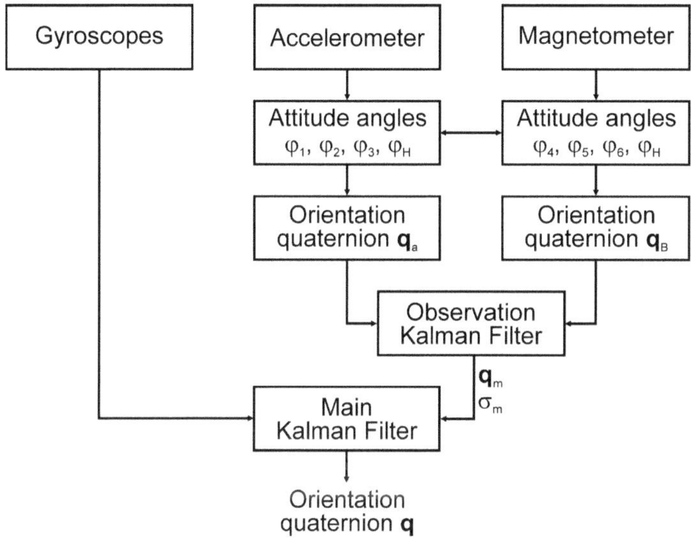

Fig. 4.12: Structure of information flow and sensor fusion in EKF.

In the main Kalman Filter, the current state is propagated only with external input directly from the gyroscopes. The measurement update is more complicated since it involves two different measurements, the state derived mainly from accelerations and mainly from the magnetic field. Both sensory paths are, with slight interactions, the basis for the derivation of a complete orientation quaternion, $\mathbf{q}_a$ and $\mathbf{q}_B$. An observation Kalman Filter, which averages with variable weights, provides the measurement quaternion for the main filter. In addition, it provides a statement on the reliability of the observation, expressed as measurement variance $\boldsymbol{\sigma}_m$.

For clarity reasons, the algorithm derived for the main Kalman Filter is described first, the measurement fusion in the observation filter can then be reduced to a subset of the same equations. A visualization of the algorithm behind equations 4.32-4.43 can be found in Fig. 4.13.

In order to cope with time varying biases and to output corrected angular rates, the inertial system state is extended to

$$\mathbf{x} = \begin{pmatrix} q_1 & q_2 & q_3 & q_4 & {}_B\omega_x & {}_B\omega_y & {}_B\omega_z & b_{\omega x} & b_{\omega y} & b_{\omega z} \end{pmatrix}^T \tag{4.31}$$

including the orientation quaternion $\mathbf{q}$, angular rates in the body-fixed system ${}_B\boldsymbol{\omega}$ and possibly time varying biases $\mathbf{b}_\omega$ as of Eq. 3.4. In the prediction step (denoted with a superscript '-') of the Extended Kalman Filter, the best estimate of the state $\mathbf{x}^+(t_i)$ (posterior: '+') is propagated in time

$$\mathbf{x}^-(t_{i+1}) = \boldsymbol{\Phi}(t_{i+1}, t_i)\mathbf{x}^+(t_i) + \mathbf{B}^+(t_i)\mathbf{u}^+(t_i) \tag{4.32}$$

with the transition matrix $\boldsymbol{\Phi}(t_{i+1}, t_i)$ and system input. For setting up the transition matrix, as well as the input matrix $\mathbf{B}^+(t_i)$, the basic equation for quaternion propagation is required [88]:

$$\begin{bmatrix} \dot{q}_1 \\ \dot{q}_2 \\ \dot{q}_3 \\ \dot{q}_4 \end{bmatrix} = \frac{1}{2} \underbrace{\begin{bmatrix} q_4 & -q_3 & q_2 \\ q_3 & q_4 & -q_1 \\ -q_2 & q_1 & q_4 \\ -q_1 & -q_2 & -q_3 \end{bmatrix}}_{{}_B\mathbf{J}_q} \begin{bmatrix} {}_B\omega_x \\ {}_B\omega_y \\ {}_B\omega_z \end{bmatrix} \tag{4.33}$$

The transition matrix

$$\boldsymbol{\Phi}(t_{i+1}, t_i) = \begin{bmatrix} \mathbf{E}^{4x4} & \mathbf{0}^{3x3} & {}_B\mathbf{J}_q \cdot \Delta t \\ \mathbf{0}^{3x4} & \mathbf{0}^{3x3} & \mathbf{E}^{3x3} \\ \mathbf{0}^{3x4} & \mathbf{0}^{3x3} & \mathbf{E}^{3x3} \end{bmatrix} \tag{4.34}$$

itself keeps the orientation quaternion and compensates for the bias with the entries in the third column of Eq. 4.34. With Δt being the time step of the discrete control algorithm, $\boldsymbol{\Phi}(t_{i+1}, t_i)$ is a first order approximation (e.g. the truncation of the Taylor expansion after the first derivative). The innovation comes from the system input

$$\mathbf{u}^+(t_i) = \begin{bmatrix} {}_B\omega_x \\ {}_B\omega_y \\ {}_B\omega_z \end{bmatrix}, \tag{4.35}$$

premultiplied with the input matrix

$$\mathbf{B}^+(t_i) = \begin{bmatrix} {}_B\mathbf{J}_q \cdot \Delta t \\ \mathbf{E}^{3x3} \\ \mathbf{0}^{3x3} \end{bmatrix}. \tag{4.36}$$

To proceed from one time step to the next, not only the state, but also the covariance matrix has to be propagated. This is done by matrix multiplication with the transition matrix and adding of the additional covariance $\mathbf{Q}(t_i)$ representing system uncertainties. This second term in Eq. 4.37 is the basis of increased uncertainty visualized with a wider Gauss curve in Fig. 4.11:

$$\mathbf{P}^-(t_{i+1}) = \mathbf{\Phi}(t_{i+1}, t_i)\mathbf{P}^+(t_i)\mathbf{\Phi}^T(t_{i+1}, t_i) + \mathbf{Q}(t_i) \tag{4.37}$$

As a last step in the prediction of the Extended Kalman Filter, the predicted prior state $\mathbf{x}^-(t_{i+1})$ is transformed to the predicted measurement $\hat{\mathbf{z}}(t_{i+1})$ for comparison with the actual measurement at timestep t_{i+1}

$$\hat{\mathbf{z}}(t_{i+1}) = \mathbf{H}(t_{i+1})\mathbf{x}^-(t_{i+1}) \tag{4.38}$$

with the output matrix

$$\mathbf{H}(t_{i+1}) = \begin{bmatrix} \mathbf{E}^{3x3} & \mathbf{0}^{3x3} & \mathbf{0}^{3x3} \end{bmatrix} \tag{4.39}$$

The correction step then updates the prior estimate with the current measurement, calculating the difference $\mathbf{v}(t_{i+1})$ between prediction and actual measurement vectors

$$\mathbf{v}(t_{i+1}) = \mathbf{z}(t_{i+1}) - \hat{\mathbf{z}}(t_{i+1}) \tag{4.40}$$

The actual measurement vector $\mathbf{z}(t_{i+1}) = \mathbf{q}_m = \begin{bmatrix} q_{1m} & q_{2m} & q_{3m} & q_{4m} \end{bmatrix}$ is the output of the observation Kalman Filter (Fig. 4.12), combining information from acceleration and magnetic field sensors. To form the final (posterior) estimate of the state at time t_{i+1}, the measurement difference of Eq. 4.40 is added to the prior estimate, premultiplied with the Kalman Gain $\mathbf{K}(t_{i+1})$

$$\mathbf{x}^+(t_{i+1}) = \mathbf{x}^-(t_{i+1}) + \mathbf{K}(t_{i+1})\mathbf{v}(t_{i+1}) \tag{4.41}$$

Obviously, the calculation of the Kalman Gain $\mathbf{K}(t_{i+1})$ is one of the special characteristics of this type of filter. Due to the structure of Eq. 4.42, it takes into account the relation between the covariance matrix $\mathbf{P}^-(t_{i+1})$, which expresses the confidence in the current estimation of the state $\mathbf{x}^-(t_{i+1})$, and the reliability of the observation with covariance matrix $\mathbf{R}(t_{i+1})$.

$$\mathbf{K}(t_{i+1}) = \mathbf{P}^-(t_{i+1})\mathbf{H}^T(t_{i+1}) \left[\mathbf{H}(t_{i+1})\mathbf{P}^-(t_{i+1})\mathbf{H}^T(t_{i+1}) + \mathbf{R}(t_{i+1})\right]^{-1} \tag{4.42}$$

In addition to the state update, its covariance matrix has to be updated. The additional information introduced by the state observation reduces uncertainties (e.g. the state covariance decreases). With the matrix product in Eq. 4.43 being positive definite, this becomes obvious in

$$\mathbf{P}^+(t_{i+1}) = \mathbf{P}^-(t_{i+1}) - \mathbf{K}(t_{i+1})\mathbf{H}(t_{i+1})\mathbf{P}^-(t_{i+1}). \tag{4.43}$$

This equation terminates the repetitive cycle of prediction and correction of the Extended Kalman Filter. In general, the initialization of the filter requires good engineering judgment [98]. For state initialization, averaging over several measurement cycles is appropriate to determine initial orientation and angular rates. The same can be done for state covariance. For the determination of the covariance matrix

$$\mathbf{Q}^{+}(t_0) = \mathbf{Q} = diag\left(\sigma_{q1}^2 \quad \sigma_{q2}^2 \quad \sigma_{q3}^2 \quad \sigma_{q4}^2 \quad \sigma_{\omega x}^2 \quad \sigma_{\omega y}^2 \quad \sigma_{\omega z}^2 \quad \sigma_{b_{\omega x}}^2 \quad \sigma_{b_{\omega y}}^2 \quad \sigma_{b_{\omega z}}^2\right), \quad (4.44)$$

the variance of each entry x_i of the state vector can be calculated with the standard formula

$$\sigma_{x_i}^2 = \frac{1}{n-1}\sum_{j=1}^{n}(x_{i_j} - x_{i_0})^2 \qquad (4.45)$$

taking the first n samples into account. For the variance of quaternion elements, this should be done on the level of the rotation angle around the Euler axis. The quality of the estimation improves if the inertial measurement unit does not move during initialization. The covariance matrix of the measurement, $\mathbf{R}$, is provided by the observation Kalman Filter. Fig. 4.13 summarizes the main Kalman Filter with a flow chart.

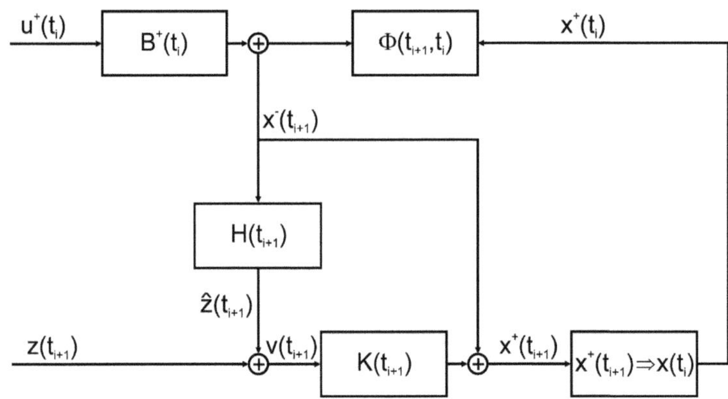

Fig. 4.13: Structure of the main Extended Kalman Filter.

In the observation filter, both measurement types are fused with respect to their reliability. In order to use the same algorithmic framework as described above, the transition matrix is set to zero, leading to a new averaging in every timestep without taking into account prior measurements. The prediction equations are thus reduced to

$$\begin{aligned}
\mathbf{x}^{-}(t_{i+1}) &= \mathbf{B}^{+}(t_{i+1})\mathbf{u}^{+}(t_{i+1}) = \mathbf{u}^{+}(t_{i+1}) \\
\hat{\mathbf{z}}(t_{i+1}) &= \mathbf{H}(t_{i+1})\mathbf{x}^{-}(t_{i+1}) = \mathbf{x}^{-}(t_{i+1}),
\end{aligned} \qquad (4.46)$$

in contrast to the previous Kalman Filter using only the latest information (t_{i+1}) and defining the system input to be the quaternion mainly derived from magnetic field measurement $\mathbf{u}^+(t_i) = \mathbf{q}_B$, the update input to be the quaternion mainly derived from accelerations $\mathbf{z}(t_{i+1}) = \mathbf{q}_a$ and the state as the final orientation quaternion $\mathbf{x}(t_{i+1}) = \mathbf{q}_m$. All other equations are reduced accordingly.

The important aspect of the observation filter is in the definition of the covariance matrices. Both measured values, accelerations and magnetic field, are subject to significant, mostly short-term, disturbances like shocks and fields emitted by electric motors. To cope with these disturbances and to make the orientation determination more robust to external influences, the observation filter detects abnormalities and adjusts the elements of the covariance matrices

$$
\begin{aligned}
\mathbf{Q}(t_{i+1}) &= diag \begin{pmatrix} \sigma_{qm} & \sigma_{qm} & \sigma_{qm} & \sigma_{qm} \end{pmatrix} \\
\mathbf{R}(t_{i+1}) &= diag \begin{pmatrix} \sigma_{qa} & \sigma_{qa} & \sigma_{qa} & \sigma_{qa} \end{pmatrix}
\end{aligned}
\tag{4.47}
$$

For the acceleration quaternion, this is achieved by increasing the variances in accordance with the relation of total accelerations $\mathbf{a}$ and gravity $\mathbf{g}$:

$$
\sigma_{qa} = \sigma_{a0} + \sigma_a \cdot \frac{|\mathbf{a}| - |\mathbf{g}|}{|\mathbf{g}|}
\tag{4.48}
$$

Whenever there are higher or lower accelerations than gravity, an increased variance signifies the lower reliability of acceleration measurement, also in comparison to magnetic field measurement. The use of absolute values of acceleration and gravity has the advantage of not requiring orientation data from the final sensor fusion, thus allowing for implementation of the algorithm on a local microcontroller, independent of the central CPU. The equivalent idea is applied to the covariance matrix of magnetic field measurement $\mathbf{Q}$.

This local observation filter has the knowledge on the reliability of both measurements. Therefore it submits the variances $\boldsymbol{\sigma}_m$ for matrix $\mathbf{R}$ of the main Kalman Filter (Eq. 4.42, Fig. 4.12).

4.3.3 Sigma-Point Kalman Filtering

Although the Extended Kalman Filter is the most popular extension, it is not an optimal filter in the sense of the linear Kalman Filter (KF). Several other variations and enhancements to the KF have been suggested to cope with drawbacks of the EKF. Recently, JULIER and UHLMANN [79] proposed the so-called Sigma Point Kalman Filter (SPKF), based on Unscented Transformations [78]. SPKF became popular for inertial navigation when a vehicle equipped with such a system won the DARPA Grand Challenge in 2005 and mastered the special needs of fully autonomous driving.

In fact, Extended Kalman Filters propagate the system state and all relevant noise densities, which are approximated as Gaussian random variables (GRV) analytically through the first-order linearization of a nonlinear system. This can induce large errors in the posterior mean and especially the covariance of the GRV [132].

Sigma Point Filters use a different approach, deterministically drawing samples, the Sigma Points. These sampling points are then propagated through the true nonlinear system capturing the posterior mean and covariance of the GRV accurately, at least to the third order. As an advantage for some systems, this does not necessarily require the explicit calculation of the Jacobian. At the expense of a higher computational cost, more Sigma Points can still increase accuracy. In comparison to Monte Carlo Filters, where sampling points are chosen randomly, SPKF require far less CPU power.

There are many alternative types of SPKF, mainly with a varying selection scheme of the Sigma Points. For this work, the Unscented Kalman Filter (UKF) is chosen. A visual summary of the algorithm is shown in Fig. 4.14. After appropriate initialization, a set of $2L + 1$ Sigma Points $\chi_k(t_i)$ are drawn at timestep t_i, with L being the length of the state vector.

$$\chi_k(t_i) = \begin{cases} \mathbf{x}(t_i) & , k = 0 \\ \mathbf{x}(t_i) + \zeta \left(\sqrt{\mathbf{P_x}(t_i)} \right)_k & , k = 1, \ldots, L \\ \mathbf{x}(t_i) - \zeta \left(\sqrt{\mathbf{P_x}(t_i)} \right)_k & , k = L+1, \ldots, 2L \end{cases} \qquad (4.49)$$

From Eq. 4.49 it is obvious that sampling points are drawn around a state $\mathbf{x}(t_i)$ with respect to the covariance matrix and with a scaling factor ζ. $\left(\sqrt{\mathbf{P_x}(t_i)} \right)_k$ denotes the kth column of the square root of the matrix. For the calculation of the square root of $\mathbf{P_x}$, with $\left(\sqrt{\mathbf{P_x}} \right) \left(\sqrt{\mathbf{P_x}} \right)^T = \mathbf{P_x}$, *Cholesky Factorization* offers an efficient algorithm. These Sigma Points are then propagated to the next time step by the function F, which is already derived in Eq. 4.32, replacing the state vector.

$$\chi_k^*(t_{i+1}) = F\left(\chi_k(t_i); \mathbf{u}(t_i) \right), \qquad k = 0 \ldots 2L \qquad (4.50)$$

The sum of the propagated Sigma Points, with weights W_k, forms the prediction of the Kalman filter.

$$\mathbf{x}^-(t_{i+1}) = \sum_{k=0}^{2L} W_k \cdot \chi_k^*(t_{i+1}) \qquad (4.51)$$

In addition, the updated covariance matrix of the prediction at time t_{i+1} needs to be calculated:

$$\mathbf{P_x}^-(t_{i+1}) = \sum_{k=0}^{2L} W_k [\chi_k^*(t_{i+1}) - \mathbf{x}^-(t_{i+1})][\chi_k^*(t_{i+1}) - \mathbf{x}^-(t_{i+1})]^T + \mathbf{Q} \qquad (4.52)$$

After that, new Sigma Points are to be drawn to improve accuracy, Eq. 4.53. This step is not mandatory and can be skipped if computational power is restricted. In order to reduce the SP to the information required for comparison to measurable variables, $\mathbf{H}(t_{i+1})$ (as of Eq. 4.39) selects the corresponding elements, Eq. 4.54.

$$\boldsymbol{\chi}_k(t_{i+1}) = \begin{cases} \mathbf{x}(t_{i+1}) & , k = 0 \\ \mathbf{x}(t_{i+1}) + \zeta \left(\sqrt{\mathbf{P}_{\mathbf{x}}^{-}(t_{i+1})} \right)_k & , k = 1, \ldots, L \\ \mathbf{x}(t_{i+1}) - \zeta \left(\sqrt{\mathbf{P}_{\mathbf{x}}^{-}(t_{i+1})} \right)_k & , k = L+1, \ldots, 2L \end{cases} \tag{4.53}$$

$$\boldsymbol{\psi}_k(t_{i+1}) = \mathbf{H}(t_{i+1}) \cdot \boldsymbol{\chi}_k(t_{i+1}) \qquad\qquad , k = 0, \ldots, 2L \tag{4.54}$$

Finally to conclude the prediction step, a weighted sum of the reduced Sigma Points signifies the predicted measurement:

$$\hat{\mathbf{z}}(t_{i+1}) = \sum_{k=0}^{2L} W_k \cdot \boldsymbol{\psi}_k(t_{i+1}) \tag{4.55}$$

For the measurement update, the Kalman gain needs to be computed in a slightly different way. In addition to the covariance of the expected measurement, $\mathbf{P}_{zz}(t_{i+1})$, it is necessary to know the probabilistic connection between state and measurement $\mathbf{P}_{xz}(t_{i+1})$ [100].

$$\mathbf{P}_{zz}(t_{i+1}) = \sum_{k=0}^{2L} W_k \left[\boldsymbol{\psi}_k(t_{i+1}) - \hat{\mathbf{z}}(t_{i+1}) \right] \left[\boldsymbol{\psi}_k(t_{i+1}) - \hat{\mathbf{z}}(t_{i+1}) \right]^T + \mathbf{R} \tag{4.56}$$

$$\mathbf{P}_{xz}(t_{i+1}) = \sum_{k=0}^{2L} W_k \left[\boldsymbol{\chi}_k(t_{i+1}) - \mathbf{x}^{-}(t_{i+1}) \right] \left[\boldsymbol{\psi}_k(t_{i+1}) - \hat{\mathbf{z}}(t_{i+1}) \right]^T \tag{4.57}$$

Similar to the implementation of the Extended Kalman Filter, the Kalman Gain $\mathbf{K}(t_{i+1})$ represents a relationship between the trust in prediction and in observation.

$$\mathbf{K}(t_{i+1}) = \mathbf{P}_{xz}(t_{i+1}) \cdot \mathbf{P}_{zz}^{-1}(t_{i+1}) \tag{4.58}$$

Little trust in the update, expressed as a high variance in measurements, leads to a small Kalman Gain, thus reducing the impact of update equation:

$$\mathbf{x}^{+}(t_{i+1}) = \mathbf{x}^{-}(t_{i+1}) + \mathbf{K}(t_{i+1}) \left(\mathbf{z}(t_{i+1}) - \hat{\mathbf{z}}(t_{i+1}) \right) \tag{4.59}$$

Finally the new covariance matrix after the update in t_{i+1} is calculated to allow for a new drawing of Sigma Points in the next timestep.

$$\mathbf{P}_{\mathbf{x}}(t_{i+1}) = \mathbf{P}_{\mathbf{x}}^{-}(t_{i+1}) - \mathbf{K}(t_{i+1})\mathbf{P}_{yy}(t_{i+1})\mathbf{K}^T(t_{i+1}) \tag{4.60}$$

The scaling factor and the weights remain to be determined. Here, engineering judgment and tests are required. JULIER and UHLMANN [78] gave the following proposition, that has been adopted and individually adjusted by SHIN [123], METZGER [100] and ZHANG [146]

$$\zeta = \sqrt{L + \lambda}, \qquad W_0 = \frac{\lambda}{L + \lambda}, \qquad W_k = \frac{1}{2(L + \lambda)} \tag{4.61}$$

For a complete overview on the parameterization please refer to [147]. Fig. 4.14 shows the structure and information dependencies in the Sigma Point Kalman Filter.

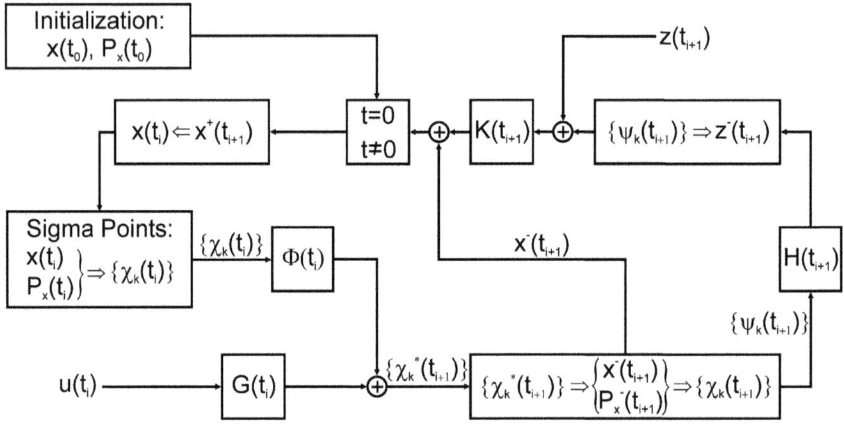

Fig. 4.14: Structure of the Sigma Point Kalman Filter.

4.3.4 Particle Filter

The two probabilistic approaches in the previous sections are both non-linear approximations for Markov processes with the assumption of Gaussian distribution of probability density functions (PDF). However with non-linear systems the transform of a Gaussian PDF becomes non-Gaussian. In order to study the influence of these effects and to see whether they can be neglected, a Particle Filter (PF) is developed for the comparison with EKF and SPKF.

Particle Filters, especially the used bootstrap filters, are an efficient form of sequential Monte Carlo Methods (SMC). In contrast to the somewhat forceful approach of standard Monte Carlo filtering, where a huge number of samples are randomly distributed, propagated and used to approximate the posterior PDF, a limited number of particles are efficiently placed at important (e.g. non zero) places of PDF in state space. The posterior is not evaluated as a function, like in EKF, but rather as weighted average of all particles, allowing for probability distributions with several maxima.

Based on the same principle of Bayesian estimation, Particle Filter also consist of a prediction and an update step. The aim of the filtering process is to obtain the estimation of state $\mathbf{x}_{i+1}$ based on the set of all available measurements $\{\mathbf{z}_j : j = 1, \ldots, i+1\}$, e.g. the probability density $p(\mathbf{x}_{i+1}|\mathbf{z}_{1:i+1})$. This distribution can be determined by the recursion: *Prediction:*

$$p(\mathbf{x}_{i+1}|\mathbf{z}_{1:i}) = \int p(\mathbf{x}_{i+1}|\mathbf{x}_i)p(\mathbf{x}_{i+1}|\mathbf{z}_{1:i})d\mathbf{x}_i \tag{4.62}$$

Update:

$$p(\mathbf{x}_{i+1}|\mathbf{z}_{1:i+1}) = \frac{p(\mathbf{z}_{i+1}|\mathbf{x}_{i+1})p(\mathbf{x}_{i+1}|\mathbf{z}_{1:i})}{\int p(\mathbf{z}_{i+1}|\mathbf{x}_{i+1})p(\mathbf{x}_{i+1}|\mathbf{z}_{1:i})d\mathbf{x}_{i+1}} \tag{4.63}$$

Note that for clarity the timestep i has been shifted to the subscript and superscripts signifying prediction and update have been omitted. For most dynamic systems, expressions and recursions cannot be calculated analytically. Thus approximations are necessary [61]. Unfortunately, it is usually not possible to sample directly from the posterior distribution [116]. Therefore, samples are drawn from an arbitrary so-called "importance sampling distribution" $\pi(\mathbf{x}_{0:i+1}|\mathbf{z}_{1:i+1})$ that is similar to $p(\cdot)$ (e.g. having the same support)

$$\pi(\mathbf{x}) > 0 \Rightarrow q(\mathbf{x}) > 0 \quad \forall \ \mathbf{x} \in \mathbb{R}^L, \tag{4.64}$$

adjusted with the importance weights

$$\tilde{w}(\mathbf{x}_{0:i+1}) = \frac{p(\mathbf{x}_{0:i+1}|\mathbf{z}_{1:i+1})}{\pi(\mathbf{x}_{0:i+1}|\mathbf{z}_{1:i+1})} \tag{4.65}$$

Assuming a computation of $p(\mathbf{x}_{i+1}|\mathbf{z}_{1:i+1})$ without modification of past results, the importance sampling distribution can be iteratively computed with

$$\pi(\mathbf{x}_{0:i+1}|\mathbf{z}_{1:i+1}) = \pi(\mathbf{x}_{0:i}|\mathbf{z}_{1:i})\pi(\mathbf{x}_{i+1}|\mathbf{x}_{i+1}, \mathbf{z}_{1:i+1}) \tag{4.66}$$

The importance weights can then be recursively calculated with [34]

$$\tilde{w}_{i+1}^{(k)} \propto \tilde{w}_i^{(k)} \frac{p\left(\mathbf{z}_{i+1}|\mathbf{x}_{i+1}^{(k)}\right) p\left(\mathbf{x}_{i+1}^{(k)}|\mathbf{x}_i^{(k)}\right)}{\pi\left(\mathbf{x}_{i+1}^{(k)}|\mathbf{x}_{0:i}^{(k)}, \mathbf{z}_{1:i+1}\right)}, \tag{4.67}$$

taking into account the Markov characteristics of the process. Now the choice of the importance sampling distribution becomes critical. In Bootstrap filters, a very convenient solution [90]

$$\pi(\mathbf{x}_{i+1}|\mathbf{x}_{0:i}, \mathbf{z}_{1:i+1}) = p(\mathbf{x}_{i+1}|\mathbf{x}_i) \tag{4.68}$$

leads to a simplification of the recursive weights estimation

$$\tilde{w}_{i+1}^{(k)} \propto \tilde{w}_i^{(k)} \cdot p\left(\mathbf{z}_{i+1}|\mathbf{x}_{i+1}^{(k)}\right) \tag{4.69}$$

As the last step of the update cycle, the importance weights need to be normalized in order to fulfill the requirements for probability densities.

$$w_{i+1}^{(k)} = \frac{\tilde{w}_{i+1}^{(k)}}{\sum_{j=1}^{N} \tilde{w}_{i+1}^{(j)}} \tag{4.70}$$

One difficulty with Particle Filter shall just be mentioned here. Due to the structure of the algorithm, after a certain number of iterations, the normalized weights of all but very few particles will be negligible. Therefore a resampling is required, for example if the effective sample size N_{eff}

$$\tilde{N}_{eff} = \frac{1}{\sum_{j=1}^{N} \left(w_i^{(j)}\right)^2} \tag{4.71}$$

is small and shows a severe degeneracy problem. Then usually all particles are split into equally weighted samples, with the number of children referring to the original weight of each particle.

For the multi-sensor fusion of the inertial measurement unit, the well known equations

$$\mathbf{x}_{i+1} = \mathbf{\Phi}_{i+1,i} \cdot \mathbf{x}_i + \mathbf{B}_i \cdot \mathbf{u}_i + \mathbf{w}_i \tag{4.72}$$

$$\mathbf{z}_{i+1} = \mathbf{H}_{i+1} \cdot \mathbf{x}_{i+1} + \mathbf{v}_{i+1} \tag{4.73}$$

are still valid. In order to keep computational costs reasonable, state and measurement vectors are reduced to represent the orientation quaternion only. With the system model as of Eq. 4.72, the particles of timestep i are propagated adding uncertainty due to the noise vector $\mathbf{w}_i$. Thus the prediction step is similar to the Kalman Filters. The significant difference comes with the calculation of $p\left(\mathbf{z}_{i+1}|\mathbf{x}_{i+1}^{(k)}\right)$ which is required for Eq. 4.69. With the measurement matrix $\mathbf{H}$ being an 4×4 identity matrix and assuming the measurement noise to be Gaussian distributed with zero mean and standard deviation σ, i.e.

$$p(\mathbf{v}) = \frac{1}{\sigma\sqrt{2\pi}} \cdot e^{-\frac{\mathbf{v}^T\mathbf{v}}{2\sigma^2}}, \tag{4.74}$$

the distribution function for the importance weight update in Eq. 4.69

$$p(\mathbf{z}_{i+1}|\mathbf{x}_{i+1}^{(k)}) = \frac{1}{\sigma\sqrt{2\pi}} \cdot exp\left(-\frac{\left(\mathbf{z}_{i+1} - \mathbf{x}_{i+1}^{(k)}\right)^T \left(\mathbf{z}_{i+1} - \mathbf{x}_{i+1}^{(k)}\right)}{2\sigma^2}\right) \tag{4.75}$$

signifies the probability distribution to get a measurement $\mathbf{z}_{i+1}$ under the condition of each particle $\mathbf{x}_{i+1}^{(k)}$. Since weights are normalized with Eq. 4.70, the first fraction of Eq. 4.75 can be ignored. Substituting in Eq. 4.69 leads to

$$\tilde{w}_{i+1}^{(k)} \propto \tilde{w}_i^{(k)} \cdot exp\left(-\frac{\left(\mathbf{z}_{i+1} - \mathbf{x}_{i+1}^{(k)}\right)^T \left(\mathbf{z}_{i+1} - \mathbf{x}_{i+1}^{(k)}\right)}{2\sigma^2}\right), \tag{4.76}$$

which are the weights of all particles $\mathbf{x}_{i+1}^{(k)}$ if a measurement $\mathbf{z}_{i+1}$ was observed. These weights are equivalent to the probability of each particle. In order to finally calculate the output of the filter, the particles need to be summed according to their normalized weights:

$$\mathbf{x}_{i+1} = \sum_{j=1}^{N} w_{i+1}^{(k)} \cdot \mathbf{x}_{i+1}^{(k)} \tag{4.77}$$

The performance of the Particle Filter depends on the parameter setting, especially on the number of particles N. Like in many cases, a tradeoff between accuracy and computational cost must be found. Therefore the algorithm is evaluated, along with the two versions of the Kalman Filter, with suitable tests.

4.4 Performance Evaluation of Stochastic Methods

4.4.1 Reference System and Test Setup

For the performance evaluation of the three stochastic multi-sensor fusion algorithms, a parallel kinematic robot moved a magnetometer equipped μCube in all six degrees of freedom. On the one hand the position and orientation of the IMU in space is determined with encoder measurements and the inverse kinematics of the robot. On the other hand the time synchronized inertial sensor data is then the basis for spatial orientation determination with the above developed algorithms. Sensor data was logged during the experiments at a frame rate of $1kHz$. For use with mathematical optimization, the actual orientation calculation was shifted to post processing. The set-up was equivalent to the tests shown in Fig. 5.6.

Due to the linear motion of the sensor, as well as vibrations originating in the joints of the robot, disturbing influences of real-world applications are included in the test results. Especially challenging are the magnetic fields emitted by the six motors of the robot, being in the close vicinity of the sensors.

4.4.2 Parameter Optimization with Implicit Filtering

The performance of all algorithms depends in part significantly on various parameters. Although most of them can be estimated within a certain range, mathematical optimization can significantly improve the final accuracy.

Due to manufacturing tolerances, the exact orientation of the sensitive axis of each sensor is not known to an accuracy required for high performance sensor fusion. A preliminary identification procedure could compensate for uncertainties within the IMU, however, does not take the exact mounting on the robot into account. Therefore the orientation of each sensor is considered as a variable parameter within tight tolerances, along with the sensitivity, which manufacturers only guarantee to up to 10% from nominal values for non-selected MEMS sensors.

The three fusion algorithms themselves mainly depend on the level of confidence attributed to each sensor type, expressed as their standard deviation. To find optimal values for these matrices they are included in the set of parameters for optimization.

The same non-linear optimization algorithm IFFCO that is also used in Section 3.8.2 is suitable for this task. Critical in this case is the formulation of the error function f, which is to be minimized. In the test setup the output of the sensor fusion algorithm corresponds to pitch (ψ), yaw (ϕ) and roll (γ). Around neutral position, the yaw angle reflects a rotation about the vertical axis, thus depending to a high extent on the magnetometer. Since the measurement of the Earth's magnetic field is in general less accurate than acceleration sensing, which is more used for pitch and roll, the yaw angle tends to significantly contribute to the error function. Therefore the error function Eq. 4.78 attributes different weights χ_ψ, χ_ϕ and χ_γ to the three output angles.

$$f = \chi_\psi \cdot \sum_{\tau=0}^{30s} (\psi_{SF} - \psi_{ref})^2 + \chi_\phi \cdot \sum_{\tau=0}^{30s} (\phi_{SF} - \phi_{ref})^2 + \chi_\gamma \cdot \sum_{\tau=0}^{30s} (\gamma_{SF} - \gamma_{ref})^2 \quad (4.78)$$

Based on this error function, the parameters of the sensor fusion algorithms are optimized, starting with nominal sensor sensitivity and optimal sensor orientation. The results are split into two groups:

- For the general case in which all orientation angles are equally relevant, equal weights are attributed to all output errors, $\chi_\psi = \chi_\phi = \chi_\gamma = 1$.

- In many cases of stabilization tasks, accuracy in pitch and roll is more important. Therefore a second optimization cycle with reduced yaw error weight $\chi_\phi = 0.25$ is aiming for higher accuracy in pitch and roll, at the expense of a reduced yaw accuracy.

Fig. 4.15 shows the optimization history of the error function for an Extended Kalman Filter algorithm. During 2000 optimization cycles, the accumulated error could be significantly reduced. With certain parameter combinations (cycle 50, 650, 930) a significant reduction could be achieved changing more relevant parameters, otherwise the error decreased gradually.

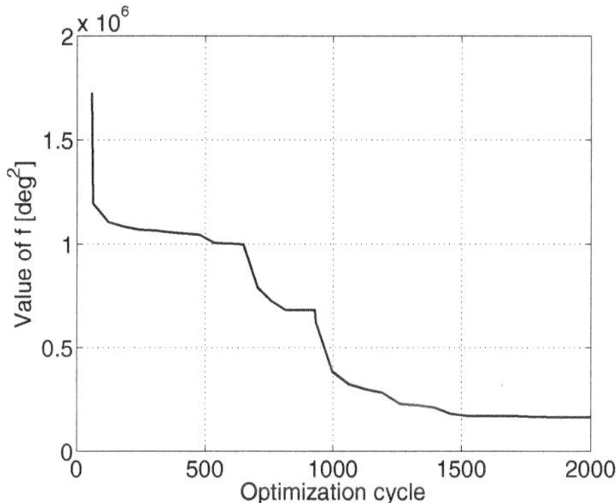

Fig. 4.15: Optimization history of the error value for the Extended Kalman Filter.

4.4.3 Accuracy of Orientation Determination

For the complete evaluation of the sensor fusion algorithms, not only the accuracy, but also the computational expense that is required needs to be considered. This is especially the case for the particle filter algorithm, which significantly depends on the number of particles used. With more particles accuracy improves, however the processing time increases so significantly that real time hardware requires a lower output rate. As a trade-off, 1500 particles are used for this analysis, representing the reasonable limit for current single board dSPACE hardware.

To perform the runtime evaluation, a 2.0 GHz processor calculates the orientation based on actual sensor measurements. Table 4.1 compares the required cycle time, normalized to the value for the Extended Kalman Filter, representing the currently used methods. The Sigma-Point Kalman Filter requires about four times the computational power, but is still in the same range as the EKF. The Particle Filter, however, requires about 13 times more time for only 1500 particles.

Table 4.1: Comparison of computational requirements for different sensor fusion algorithms, normalized to the value for the Extended Kalman Filter.

Algorithm	Computational expense [%]
EKF	100
SPKF	420
Particle Filter	1350

The accuracy evaluation is based on a $30s$ sequence of motion in all six degrees of freedom on a robot with parallel kinematic. Fig. 4.16 plots the orientation of the robot platform derived from inverse kinematics as a reference in addition to the sensor fusion result of the EKF. The left graph for the pitch angle is based on optimization with a reduced weights for yaw, Section 4.4.2. The graph for the yaw angle results in attributing equal weights for all orientation angles. If sufficient computational power is available, two algorithms with the two parameter sets can run in parallel, one for the calculation of pitch and roll, the other for the output of yaw.

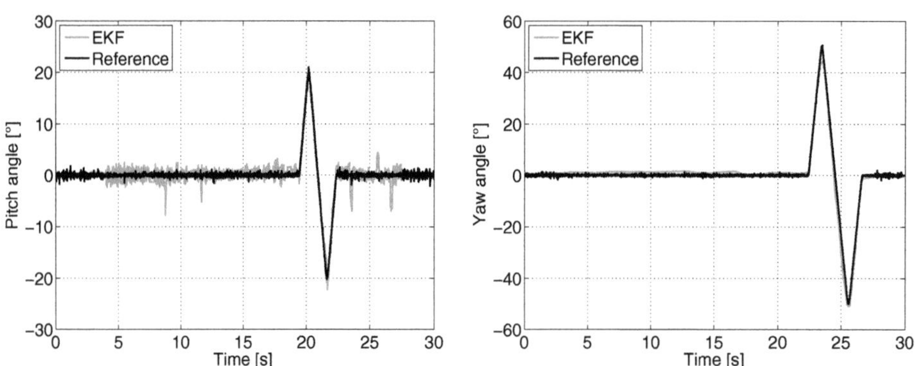

Fig. 4.16: Comparison of pitch (*left*) and yaw (*right*) angle calculated with inertial sensor fusion and robot orientation.

Attribution of equal weights leads to accuracy results of Table 4.2. For each sensor fusion algorithm, the value of the RMS error is presented with and without mathematical optimization. Note that a higher accuracy in yaw direction after optimization can result in lower performance for pitch and roll due to the attribution of sensor weights. For all algorithms the final accuracy in pitch and roll is around 2°. In yaw direction, the spread is from 2°to 11°. The lower performance of the Particle Filter indicates a higher required number of particles to demonstrate superior performance. With computational power increasing, this will be possible in the future.

With the domination of yaw error, the Sigma-Point Kalman Filter cannot show superior performance.

Table 4.2: Accuracy (in degree) for pitch, yaw and roll with equal weight distribution for mathematical optimization.

	pitch		yaw		roll	
	w. opt.	w/o opt.	w. opt.	w/o opt.	w. opt.	w/o opt.
EKF	2.03	1.82	2.27	14.60	1.40	0.94
SPKF	1,57	2.67	6.95	18.95	2.85	2.09
PF	2,32	5,22	10,87	13,68	2,85	5,41

When the importance of the yaw angle error is reduced for optimization with an error weight of $\chi_\phi = 0.25$, the accuracy in pitch and yaw becomes significantly better, Table 4.3. The rms error in pitch and roll decreases to about 1°. In this case, the Sigma-Point Kalman Filter proves superior performance by more than 0.1°

Table 4.3: Accuracy (in degree) with the focus on pitch and roll angle.

	pitch		yaw		roll	
	w. opt.	w/o opt.	w. opt.	w/o opt.	w. opt.	w/o opt.
EKF	1.29	1.82	11.50	14.60	1.03	0.94
SPKF	1.18	2.67	11.91	18.95	0.88	2.09
PF	3,28	5,22	11,18	13,68	2,98	5,41

In general, the results for the Sigma-Point Kalman Filter suggest the use of this algorithm for real-time applications since its computational demands can still be fulfilled with standard hardware. For the application of Particle Filters a high number of particles is required. Therefore it is better suited for offline applications.

5 Active Camera Stabilization

5.1 Motivation

Besides determination of orientation in space, there exist further applications of inertial measurement units. Active camera stabilization is among the commercially most relevant ones. In the automotive industry, only car fixed cameras are used for driver assistance systems up until now. Pivotable cameras can provide additional relevant information for evaluation of the environment. In aviation, inertially stabilized vision platforms are already used for ground surveillance systems. High precision hardware at even higher costs provides the required quality of sensory signals and actuators.

In contrast to existing systems, the approach of this work is to use relatively cheap hardware. To cope with deficits and to avoid time consuming, costly and for MEMS sensors not always satisfying calibration, an adaptive algorithm is developed. The idea is to install a camera stabilization system with nominal characteristics setting, which leads to imperfect results in the beginning. After a reasonable time, adaptation optimizes the outcome. Of course, complete adaptation does not need to be performed every time at startup, but previous settings are stored. However, adjustments are continuously made to account for example for changes due to deterioration.

Further industrial applications besides driver or pilot assistance include, for example, high precision robots. A stabilized camera mounted on the end effector can provide vision for a higher accuracy in assembly or tooling processes.

As mentioned in Chapter 2, nature has solved the problem of gaze stabilization in the human body very successfully. Knowledge derived from neuroscience will be used in this chapter. The first part will be a preparation for the camera stabilization used for the driver and pilot assistance systems in Chapter 6, in which an external source provides a line of sight for the stabilized camera. Section 5.3 then focuses on a closed-loop system, in which the camera image itself provides the desired line of sight. Since especially this part is bioanalogue, neuroscience plays an important role for modeling.

W. Günthner: *Enhancing Cognitive Assistance Systems with Inertial Measurement Units*, Studies in Computational Intelligence (SCI) **105**, 79–99 (2008)
www.springerlink.com

5.2 Adaptive Control for Inertial Camera Stabilization

5.2.1 Decorrelation Control

Active camera stabilization can be regarded as form of vibration isolation. A vehicle is for example jarred by bumps in the road, the camera, which is pivotably fixed to the frame shall not be affected. In contrast to many other applications of active vibration control (AVC), disturbances in the driver or pilot assistance system are not periodic as are imbalances in rotating machines. The possibility of having a smooth road for a long time and then suddenly a bump makes many standard AVC algorithms unsuitable. A second problem is superimposed intended gaze shifts and pursuit motion, which require special attention in order to not affect adaptation.

Basically, the task can be seen as a form of interference cancellation. The camera is meant to move on reference trajectory s, Fig. 5.1. Additional disturbances n lead to undesired camera motion $s + n$. An inertial measurement unit provides an estimate of the disturbance $\tilde{n}$, which is not necessarily correct in magnitude and phase.

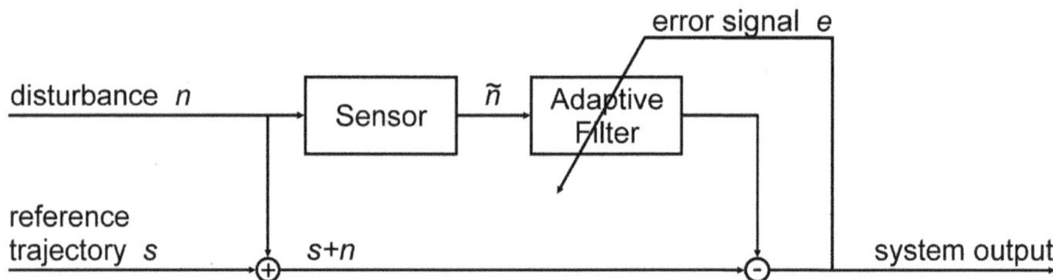

Fig. 5.1: Block diagram of adaptive interference cancellation in a 1 DOF system.

The adaptive filter then uses these estimates to calculate a filter output that shall equal the actual disturbance n as a motor command to compensate for the disturbance. The system output is used as an error signal to teach the adaptive filter.

The constitution of the error signal and the learning rule lead to the designation *decorrelation control*. This becomes visible in Fig. 5.2, where the intended line of sight is kept constant, which equals setting the reference trajectory input to zero, and including the hardware with plant and camera. A disturbing motion input on the left side is sensed by the IMU and decomposed into different variations of the sensor signals. This could be filtered signals, derivations, tapped delay lines, etc. To symbolize that, several lines connect the decomposition block and the decorrelator. The unaltered sensor signal is used on the direct nominal compensation pathway to generate a primary stabilization command for the camera motion device (plant). The difference between actual motion and stabilization is sensed by the camera since imperfect stabilization generates an optic flow (OF) on the CCD chip of the camera. This optic feedback is used as an error signal for the adaptive decorrelator. In fact, the information flow with decomposition and nominal compensation is directly inspired by gaze stabilization in the biological system, Section 2.2.

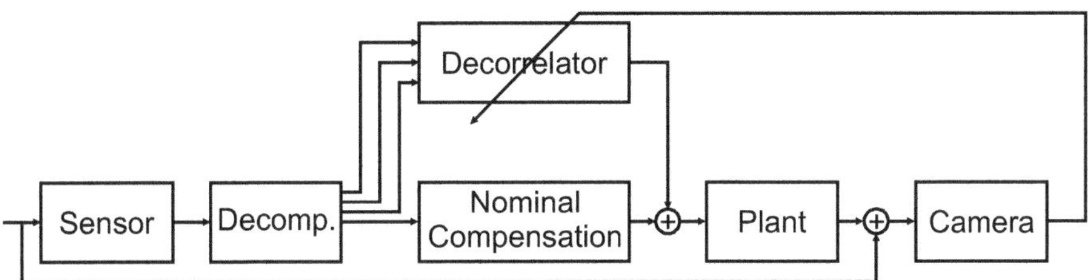

Fig. 5.2: Block diagram of the decorrelation algorithm.

The decorrelator adaptively and linearly combines all input signals to calculate the output. The weight $w_i(t)$ of input d is changed according to its correlation with the error signal, which is the optic flow OF from the camera

$$\delta w_i(t) = -\beta(t) \cdot d_i(t) \cdot OF(t) \tag{5.1}$$

and the learning rate β. By adjusting the weight according to the product of input and error signal, all correlation between them is removed. Adaptation continues until there is no dependency any more, which means the optic flow is no more related to the sensor signals of the IMU. When this is achieved, OF is not related to the vehicle's self motion but only to external influences like overtaking cars etc. Then the camera is perfectly stabilized.

If, for example, a vehicle passes in front of the car equipped with the stabilized camera, this generates optic flow. However, if the car itself does not move in a way that would generate this flow, no weight adjustment is made. The only problem arises when by accident this externally generated flow corresponds to the car's motion for a short time. But this generates misadjustment only for a very limited time. If the learning rate is small enough the effect can be neglected.

After this short general introduction to the basic ideas behind the adaptive algorithms, the concrete implementations will be detailed in the next section.

5.2.2 Adaptive LMS Algorithms

Before going into the details of the algorithm, it is useful to focus on the specific tasks adaptation has to solve and restrictions that apply. This is important in order to design a custom-fit algorithm that is computationally not too expensive to run in real-time. There are three main issues to be addressed:

- The MEMS sensors of μCube are manufactured in a highly sensitive process on a silicon wafer. In addition, the sensor die produces only a small signal, which is then amplified. Therefore, the gyroscope and accelerometer manufacturer guarantee sensor sensitivity to be within a $\pm 10\%$ range from nominal values. For camera stabilization with aperture angles of less than $10°$, this delivers unacceptable results. At angular rates of $100°/s$, this would mean a shift of more than one image per second. Thus the sensitivity (e.g., the sensor gain) must be adaptively adjusted.

- The exact and orthogonal orientation of the sensitive axes of the IMU cannot, or only with an unreasonably expensive effort, be guaranteed. Even if all sensors inside μCube are perfectly placed, the mounting of the IMU itself inside a vehicle is subject to tolerances. So if the adaptive algorithm can compensate for imperfect mounting, production tolerances can be loosened, thus additionally reducing manufacturing costs.

- The third uncertainty is in the dynamic behavior of the complete camera system. The bandwidth of the inertial measurement unit is well known and with a possible range of up to 100Hz by far high enough not to have a significant influence, especially since the first eigenfrequency of a car frame is low, about 2Hz for the Audi A8 used in test drives. The bandwidth of the FORBIAS camera motion device (CMD) by WAGNER [135] however has a bandwidth somewhere between 5 and 10Hz, depending on the exact configuration. Since it cannot be determined due to a high variety of possibilities, the adaptive algorithm has to compensate for the dynamic behavior of the CMD.

Since the control algorithm shall be portable to a PowerPC or even a small microcontroller, another consideration reduces computational demands. As mentioned in the introduction to the neuroscience approach to gaze stabilization in Section 2.2, humans have a system of translational VOR (tVOR) and rotational VOR (rVOR). This is necessary since the eyes can focus on targets just centimeters from the head in the presence of head motion. For the technical applications in connection with driver or pilot assistance systems, objects to be focused on will be at least 40-50m ahead. Assuming that translational motion perpendicular to the line of sight is restricted to the order of magnitude of a car's suspension, this results in compensatory motor commands of few hundredths of a degree. These amplitudes are in the range of the regulation error of the CMD, thus the consideration of translational motion does not improve the image quality. The same is true for the intended helicopter application. Higher frequency motion, that would require tVOR is a result of wind gusts and has the same low amplitudes. Sidewards flight of the helicopter can be accounted for with the slowly changing line of sight. Therefore, the remaining part of this book will focus on the compensation of rotation. If for any reason translation needs to be considered in some other application, the same algorithms can be expanded nearly one-to-one for tVOR.

In order to describe the developed control algorithm, the basic principle of decorrelation (Fig. 5.2) is revisited and redrawn in more technical terms, Fig. 5.3. Again, the inertial data, now restricted to the gyroscope measurements $\tilde{\omega}$, is decomposed. The combination of a nominal compensatory command n and the adaptive path a form the motor command m for the plant, e.g. the camera motion device. To be able to implement the adaptive part of the algorithm on a small microcontroller, the computationally inexpensive structure of least mean square (LMS) filters are suitable. For a theoretical review of these filters, refer to [89], [142] and [64].

The major enlargement in Fig. 5.3 are delays and phase shifts in image processing and plant, respectively. At a frame rate of 25Hz, the camera takes an image every $40ms$. Adding processing time of the software to extract optic flow from the images and additional time for transmission, optic information, which is used for adaptation can arrive more than $80ms$ later at the adaptive block W than the corresponding inertial data that caused the optic flow. In the worst case, this could lead to out-of-phase adaptation and instability of the complete algorithm. To solve this problem, two versions of the decomposed signals are sent to the adaptive path. In addition to the normal set of signals d, a second version $\tilde{d}$, which is delayed by an estimate of the timesteps Δ required for the visual path, is used for adaptation. That means that the adaptive output a is calculated directly from d in order not to artificially induce delays and only the adaptation is delayed.

The second change is similar but concerns the dynamic behavior of the plant. If an input motor command m to the plant contains parts of the frequency spectrum in the range of the plant's cutoff frequency or higher, the plant causes a phase shift (e.g., a frequency dependant delay). Thus again, the visually detected effects (OF) are no more in phase with the input to adaptation. Therefore, $\tilde{d}$ is also filtered with an estimated $\hat{P}$ of the plant's transfer function to get both inputs aligned in time.

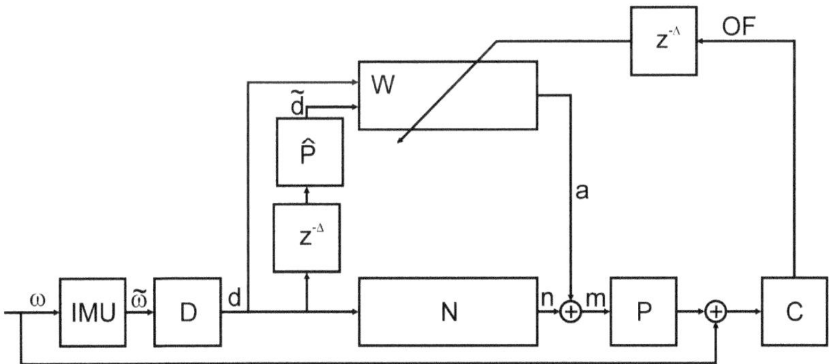

Fig. 5.3: Block diagram of the filtered-x LMS algorithm including decomposition D, nominal compensation N, plant (CMD) P, camera C and the adaptive block W.

The anti-Hebbian learning rule of Eq. 5.1 is then rewritten as

$$\delta w_i(t) = -\beta(t) \cdot (G_P \cdot d_i(t - \Delta \cdot T_s)) \cdot OF(t) \tag{5.2}$$

with the sample time T_s and the transfer function G_P of the plant. This is especially relevant at the beginning of the adaptation when the dynamic behavior of the plant has not yet been compensated for, and can be switched of later. The quality of the estimate of P is only secondary. This so-called filtered-x LMS (FxLMS, due to the filtering of the input signal) algorithm is robust to uncertainties [36], [63]. For linearization reasons, equal dynamic behavior can be assumed for each axis.

Evaluation of Fig. 5.3 allows one to make the critical decision of how the sensory signal $\tilde{\omega}$ shall be decomposed by D. An optimally stable image and, for the case of a static environment zero optic flow, would be reached if the transfer function of the sensor information propagation path equals unity, e.g.

$$G_{gyro} \cdot (G_N + G_W) \cdot G_P = 1 \tag{5.3}$$

The dynamic behavior of the gyroscopes has been derived in Eq. 3.8. The designer of the camera motion device approximates its transfer function with a single pole PT_1 system. Expanding Eq. 5.3

$$\frac{S_N}{T_B T_I s^2 + (T_B + T_I)s + 1} \cdot (G_N + G_W) \cdot \frac{1}{T_P s + 1} = 1 \tag{5.4}$$

with Laplace operator s and assuming that the sensors' time constants corresponding to cut-off frequencies of 80Hz and 400Hz have no significant influence in the range of the plant's operating frequency, Eq. 5.3 yields

$$(G_N + G_W) \cdot \frac{S_N}{T_P s + 1} \approx 1 \tag{5.5}$$

With the nominal compensation path having proportional behavior, the adaptive path must approximately show PD behavior

$$G_W \propto T_W s + 1 \tag{5.6}$$

This consideration is relevant for the type of decomposition in D. Algorithms usually implemented for active noise and vibration control decompose sensor signals into a high number of copies of these signals with different delays (e.g., a tapped delay line). Like that, a complex transfer function can be built. However, it requires permanent excitation, since constant readaptation is necessary due to the fact that leakage is necessary in the adaptation parameters in order to stabilize the algorithm. However, this cannot be guaranteed for driver assistance systems since the car might stand still for example at a traffic light or drive on a new and planar highway where car rotation does not suffice for adaptation.

Eq. 5.6 now suggests a rather simple decomposition. If the gyroscope signals are decomposed into their nominal value and their derivations in time, $d = \left(\omega \quad d\omega/dt\right)^{T}$, an adaptive linear combiner satisfies the requirements. With the derivations, the dynamic behavior of the plant can be compensated for. And since all three gyroscopes can be combined for each nominal sensitive axis, not only the sensor gain, but also the non-orthogonality and imperfect mounting of the sensors are accounted for. This can be expressed as

$$\omega_a = w_{a1} \cdot \omega_1 + w_{a2} \cdot \omega_2 + w_{a3} \cdot \omega_3, \quad a \in \{x, y, z\} \tag{5.7}$$

for each nominal axis with gyroscope 1,2 and 3. If the sensitive axes of the three gyroscopes are not coplanar and at least more or less close to orthogonal, any unknown position of the IMU can be compensated for.

Until now, only the one-dimensional case has been presented for showing the general idea behind the algorithm. For the three-dimensional case with tilt (ψ), pan (ϕ) and roll (γ) compensation, the task gets nonlinear, however can be linearized. The main difference to the principle study in Fig. 5.3 is in the position of adaptive blocks, Fig. 5.4.

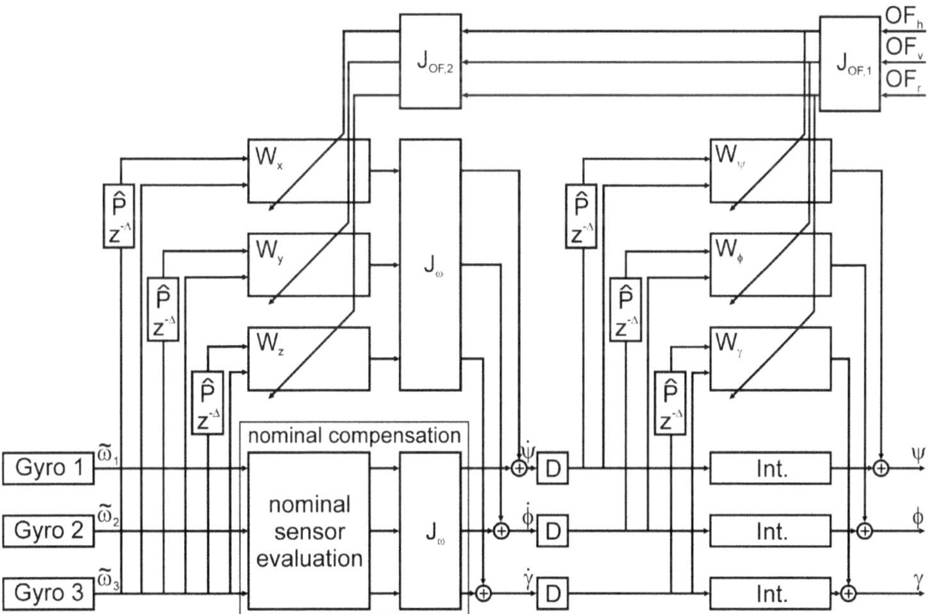

Fig. 5.4: Block diagram of three-dimensional gaze stabilization. Note that input for blocks W_x, W_y and W_z come from all three gyroscopes.

In fact, adaptation is divided into two parts. The first part (subscript p), parallel to the nominal compensation, is responsible for correcting sensor gain and compensating for non-orthogonalities according to Eq. 5.7. The second part (subscript d) adds an adaptive derivative path parallel to the integrator to the motor command to compensate for the CMD transfer behavior.

The gyroscope signals $\tilde{\omega}_1$, $\tilde{\omega}_2$ and $\tilde{\omega}_3$, that correspond approximately to x, y and z, generate a nominal compensation command $\mathbf{n}(t)$

$$
\mathbf{n}(t) = \begin{pmatrix} n_\psi(t) \\ n_\phi(t) \\ n_\gamma(t) \end{pmatrix} = \mathbf{J}_\omega(t) \cdot \mathbf{S} \cdot \begin{pmatrix} \tilde{\omega}_1(t) \\ \tilde{\omega}_2(t) \\ \tilde{\omega}_3(t) \end{pmatrix} =
$$
$$
= \frac{1}{\cos\phi} \begin{pmatrix} \cos\phi & \sin\phi\sin\psi & -\sin\phi\cos\psi \\ 0 & \cos\phi\cos\psi & \cos\phi\sin\psi \\ 0 & -\sin\psi & \cos\psi \end{pmatrix} \begin{pmatrix} S_1 & 0 & 0 \\ 0 & S_2 & 0 \\ 0 & 0 & S_3 \end{pmatrix} \begin{pmatrix} \tilde{\omega}_1(t) \\ \tilde{\omega}_2(t) \\ \tilde{\omega}_3(t) \end{pmatrix} \tag{5.8}
$$

with the current tilt, pan and roll angles of the camera motion device as measured by its encoders, nominal gyroscope sensitivities S_1, S_2 and S_3 and the gyroscope output $\tilde{\boldsymbol{\omega}}(t)$. The premultiplication with the Jacobian $\boldsymbol{J}_\omega(t)$ projects angular rates on the degrees of freedom of the camera motion device.

Parallel to the nominal path is the adaptive part compensating for the inertial measurement. Its output is calculated by

$$
\mathbf{a}_p(t) = \frac{1}{\cos\phi} \begin{pmatrix} \cos\phi & \sin\phi\sin\psi & -\sin\phi\cos\psi \\ 0 & \cos\phi\cos\psi & \cos\phi\sin\psi \\ 0 & -\sin\psi & \cos\psi \end{pmatrix} \begin{pmatrix} w_{x1} & w_{x2} & w_{x3} \\ w_{y1} & w_{y2} & w_{y3} \\ w_{z1} & w_{z2} & w_{z3} \end{pmatrix} \begin{pmatrix} \tilde{\omega}_1(t) \\ \tilde{\omega}_2(t) \\ \tilde{\omega}_3(t) \end{pmatrix} \tag{5.9}
$$

with the adaptive weight maxtrix $\mathbf{W}_p \in \mathbb{R}^{3x3}$ and completes the determination of angular rates necessary to stabilize the camera

$$
\begin{pmatrix} \dot{\psi}(t) \\ \dot{\phi}(t) \\ \dot{\gamma}(t) \end{pmatrix} = \mathbf{n}(t) + \mathbf{a}_p(t) \tag{5.10}
$$

The adaptation itself of these blocks is driven by optic flow, projected by Jacobians onto the nominal sensor axes and the gyroscope signals using the anti-Hebbian learning rule of Eq. 5.2:

$$
\begin{aligned}
\mathbf{W}_p(t) &= \mathbf{W}_p(t - T_s) - \boldsymbol{\beta}_p(t) \cdot \mathbf{J}_{OF_2}\mathbf{J}_{OF_1}\mathbf{OF}(t)\tilde{\mathbf{d}}_p^T(t) = \\
&= \mathbf{W}_p(t - T_s) - \boldsymbol{\beta}_p(t) \cdot \mathbf{A}_{SC}\mathbf{OF}(t)(g_P * \tilde{\boldsymbol{\omega}})^T(t - \Delta \cdot T_s) = \\
&= \mathbf{W}_p(t - T_s) - \boldsymbol{\beta}_p(t) \cdot \mathbf{A}_{SC} \begin{pmatrix} OF_v \\ OF_h \\ OF_r \end{pmatrix} \begin{pmatrix} g_P * \tilde{\omega}_1(t - \Delta \cdot T_s) \\ g_P * \tilde{\omega}_2(t - \Delta \cdot T_s) \\ g_P * \tilde{\omega}_3(t - \Delta \cdot T_s) \end{pmatrix}^T
\end{aligned} \tag{5.11}
$$

The learning rate $\beta_p(t)$ determines the speed of convergence of the algorithm. A_{SC}, which is the transformation matrix from the camera system to the nominal inertial sensor system, is a Cardan transformation matrix due to the cardanic structure of the camera motion device,

$$\mathbf{A}_{SC} = \begin{pmatrix} \cos\phi\cos\gamma & -\cos\phi\sin\gamma & \sin\phi \\ \cos\psi\sin\gamma + \sin\psi\sin\phi\cos\gamma & \cos\psi\cos\gamma - \sin\psi\sin\phi\sin\gamma & -\sin\psi\cos\phi \\ \sin\psi\sin\gamma - \cos\psi\sin\phi\cos\gamma & \sin\psi\cos\gamma + \cos\psi\sin\phi\sin\gamma & \cos\psi\cos\phi \end{pmatrix} \quad (5.12)$$

This projection of the error, or adaptation signal onto the nominal axis of the sensor guarantees the adaptation towards them. The "$*$" denotes the convolution of the weight function of the plant, which is the inverse Laplace transform of its transfer function

$$g_P(t) = \mathscr{L}^{-1}\left(G_P(s)\right) \quad (5.13)$$

and the angular rates delayed by $\Delta \cdot T_s$. The choice of the learning rate (LR) $\beta(t)$ is critical for the speed of adaptation, the excess error and stability of the algorithm. The latter is guaranteed for learning rates that do not exceed a critical limit [36], [142]. In practice, this was not a problem in the desired applications. So the learning rate is a trade-off between the duration of adaptation and the final mean square error (MSE). Obviously, the larger β, the further the weights cross the optimal solution after successful adaptation.

Therefore the learning rate itself is also adapted. In the diagonal matrix $\boldsymbol{\beta}_p(t) = diag[\beta_{px}(t), \beta_{py}(t), \beta_{pz}(t)]$ each LR is individually adjusted in accordance to the adaptation history. If the error has been correlated with angular rates in the same direction, e.g. with the same sign of the product, for several timesteps, the weights are still far from the optimum and the learning rate is increased. However, if the direction of adaptation often changes, the solution is close to the optimum and the LR is decreased in order to minimize the final mean square error. Or more precisely:

$$\beta_{pa}(t) = \begin{cases} \beta_{pa}(t - T_s) + \Delta\beta & if\ \mathrm{sgn}(\delta w_a(t)) = \mathrm{sgn}(\delta w_a(t - T_s)) = \mathrm{sgn}(\delta w_a(t - 2T_s)) \\ \beta_{pa}(t - T_s) - \Delta\beta & if\ \mathrm{sgn}(\delta w_a(t)) \neq \mathrm{sgn}(\delta w_a(t - T_s)) \wedge \\ & \qquad \mathrm{sgn}(\delta w_a(t - T_s)) \neq \mathrm{sgn}(\delta w_a(t - 2T_s)) \\ \beta_{pa}(t - T_s) & else \end{cases} \quad (5.14)$$

for $a \in \{x, y, z\}$. After this step, the angular rates of all axis of the camera motion device are known. Now the second superimposed adaptive path is required to compensate for the dynamic behavior of the camera motion device. As derived in

Eq. 5.6, a derivative part has to be added. In order not to use a numerically differentiated signal for the motor command, the required angular velocities are added adaptively after the integrator. To avoid drift of the CMD angles, the integrator is slightly leaky,

$$\begin{pmatrix} \psi(t) \\ \phi(t) \\ \gamma(t) \end{pmatrix} = \int_0^t \begin{pmatrix} \dot{\psi}(\tau) - \alpha \left(\psi(\tau - T_s) - \psi_r(\tau - T_s) \right) \\ \dot{\phi}(\tau) - \alpha \left(\phi(\tau - T_s) - \phi_r(\tau - T_s) \right) \\ \dot{\psi}(\gamma) - \alpha \left(\gamma(\tau - T_s) - \gamma_r(\tau - T_s) \right) \end{pmatrix} d\tau + \begin{pmatrix} a_{d\psi}(t) \\ a_{d\phi}(t) \\ a_{d\gamma}(t) \end{pmatrix} \tag{5.15}$$

and drifting to the external reference with leakage factor α. The external reference is the desired line of sight provided by image processing for the driver assistance system or the navigation computer in the helicopter application.

Decomposition in blocks D add derivations of the incoming angular rates to the signal. Since the control algorithms has to be programmed for discrete timesteps, the derivation will be carried out as a difference quotient. With a sampling time of $T_s = 1ms$, a difference over just one timestep would blow up noise too much. Therefore, an averaging over several samples, spaced by five milliseconds significantly reduces noise and only differentiates the persistent changes in the signals. For each desired motor command, the decomposition then yields

$$\mathbf{d}_{di}(t) = \begin{pmatrix} i(t) \\ D_i(t) \end{pmatrix} = \left(\frac{i(t)}{\frac{i(t)+i(t-T_s)+i(t-2T_s)-i(t-7T_s)-i(t-8T_s)-i(t-9T_s)}{21T_s}} \right), \quad i \in \{\dot{\psi}, \dot{\phi}, \dot{\gamma}\} \tag{5.16}$$

Similar to Eq 5.9, the output of the adaptive derivative path is calculated by

$$\mathbf{a}_d(t) = \begin{pmatrix} w_{\dot{\psi}} & 0 & 0 \\ 0 & w_{\dot{\phi}} & 0 \\ 0 & 0 & w_{\dot{\gamma}} \end{pmatrix} \begin{pmatrix} \dot{\psi} \\ \dot{\phi} \\ \dot{\gamma} \end{pmatrix} \tag{5.17}$$

Note that this time no cross-axis adaptation is necessary, resulting in fewer weights. For adjusting the weights, the optic flow is projected onto the degrees of freedom of the camera motion device with

$$\begin{pmatrix} OF_\psi \\ OF_\phi \\ OF_\gamma \end{pmatrix} = \mathbf{J}_{OF1} \begin{pmatrix} OF_v \\ OF_h \\ OF_r \end{pmatrix} = \frac{1}{\cos\phi} \begin{pmatrix} \cos\gamma & -\sin\gamma & 0 \\ \sin\gamma\cos\phi & \cos\gamma\cos\phi & 0 \\ -\sin\phi\cos\gamma & \sin\gamma\sin\phi & \cos\phi \end{pmatrix} \begin{pmatrix} OF_v \\ OF_h \\ OF_r \end{pmatrix} \tag{5.18}$$

To calculate the weight increments, the derivations $D_i(t)$ are now used, with the same prefiltering as before:

$$\delta w_i(t) = \beta_{d\dot{i}}(t) \cdot OF_i(t) \cdot (g_P * D_i(t - \Delta \cdot T_s)), \quad i \in \{\dot{\psi}, \dot{\phi}, \dot{\gamma}\} \tag{5.19}$$

and variable learning rates, adjusted similar to Eq. 5.14. Actually, the filtering of the input signal to adaptation can be neglected once the adaptation of W_ψ, W_ϕ and W_γ is close to the optimum since the effects of the input signal are then immediately sensed by the optical measurement, reducing the algorithm from an FxLMS to a delayed LMS structure.

Until now, one point has not yet been addressed. Image processing calculates optic flow in pixel per frame. To correctly relate this to a signal usable for adaptation in connection with angular rates, the distance to the objects in the image would be required. In practice, a mean distance is sufficient since the objective of adaptation is to bring the correlation of input and OF to zero, for which only the sign of the optic flow is relevant.

Finally, the output of the adaptive algorithm is formed by the motor command angles for the camera motion device that produce a stable image. The data rate corresponds to the control cycle of 1kHz.

To verify the efficiency of the algorithm, a simulation model for the complete 3 degree-of-freedom camera stabilization was implemented. Fig. 5.5 exemplarily shows the results for the pitch axis. An artificially lowered vestibular gain (e.g. gyroscope sensitivity), produced an optic flow of one pixel per frame that could be significantly reduced during the first 15-20 seconds. The right plot displays the behavior of the learning rate. From an initial value it increases until an upper boundary is reached. Limitation is reasonable in order to avoid unwanted effects of disturbances like the ones pointed out in Section 5.2.1.

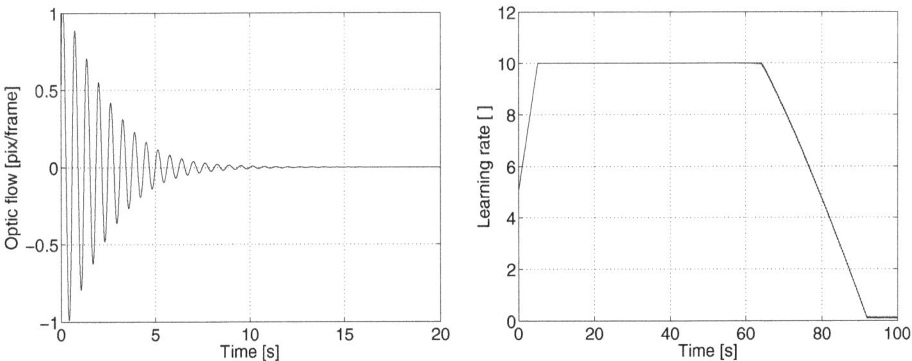

Fig. 5.5: Simulated camera stabilization with sinusoidal excitation. Optic flow corresponds to the tilt axis (*left*), learning rate to the proportional path of tilt (*right*).

After about 60 seconds, the algorithm considers the weight adjustment to be close to the optimal (e.g. the direction in which the weights are adjusted changes at every timestep), Eq. 5.14. In order to minimize the mean square error, the learning rate decreases until a lower limit is reached. A reduction to zero is not useful since a small LR can cope slowly changing demands in the weights.

5.2.3 Experimental Testing with a Steward-Gough Platform

For quantitative testing of the complete camera stabilization system, the setup is mounted on the upper platform of the Stewart-Gough platform developed at the Institute of Applied Mechanics at TU München [115], [114]. This guarantees reproducible conditions for different test scenarios. With the comparatively high power of the actuators, the platform allowed for frequency range of up to 10Hz. The inertial measurement unit was rigidly mounted next to the camera motion device, due to the structure of the adaptive sensor recombination of Eq. 5.9, no special alignment is required. Fig. 5.6 gives an overview of the test setup.

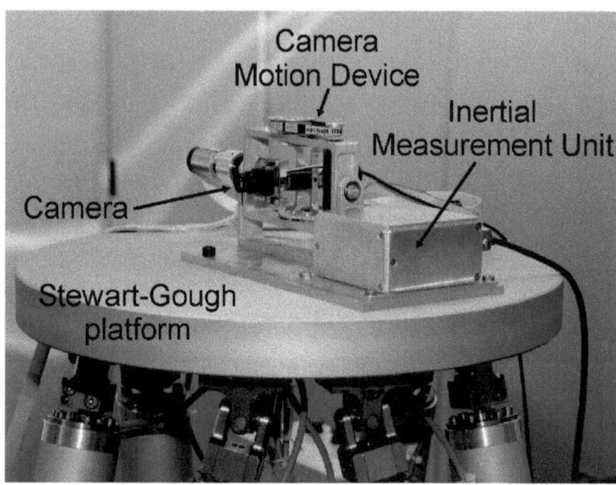

Fig. 5.6: Camera stabilization system including the inertial measurement unit mounted on the Stewart-Gough platform.

The camera motion devices used throughout this book have been designed by WAGNER [136], [138] as part of a parallel project within FORBIAS. Using a serial structure, three cardanic frames can be individually actuated. With reference to the fixation of the CMD, an 11 Watt Maxon motor (RE-max24) in combination with a backlash-free HarmonicDrive (HDUC-5-50) gearbox with reduction ratio 50:1 rotates the second frame in tilt direction. Within this frame, another HarmonicDrive servo actuator (PMA-5A-50) turns the suspension of the roll axis via a pushrod in pan direction. Finally, a CCD camera is directly mounted on the gearbox output shaft of the roll actuator. Analogue controllers (ADS) are used to control the velocity of the motors [137]. A superordinate PD position control loop receives input angles from the adaptive algorithm outlined in the previous section. Actual velocity and position from each motor are delivered by incremental encoders on each motor drive shaft with a resolution of 2048 quadcounts per revolution [60].

The camera itself is a Lechner CCTV SK-1004X, featuring a SONY EXview CCD picture element with a resolution of 500x582 pixel. Since it is an analog PAL camera, the image processing computer acquires the video signal with a frame grabber before extracting optic flow.

Fig. 5.7 represents the structure of the test setup. Note that the camera stabilization is completely independent of the hexapod control, information on the orientation of the platform is only used for recording and evaluation like in Fig. 5.8 (*left*). Due to the closed loop kinematic structure, deterministic excitation with sinusoidal motion is most convenient. For the test, amplitudes of 12.5° in each cardan angle, that describe the orientation of the platform, were achieved in the reference trajectory.

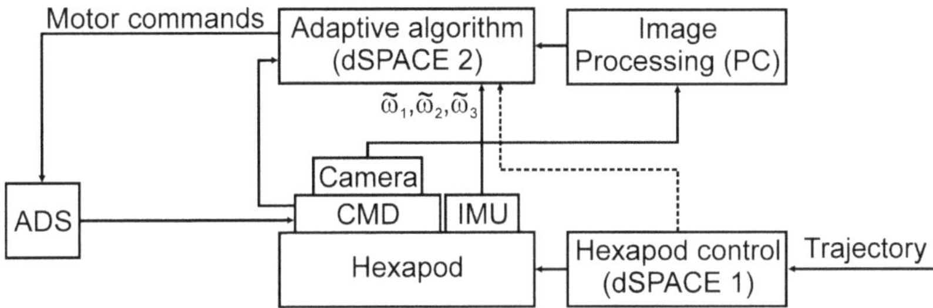

Fig. 5.7: Structure of the setup on the Stewart-Gough platform.

In addition to controlling the camera stabilization system, the dSPACE PowerPC DS1103 denoted "2" in Fig. 5.7 recorded the system state. Fig. 5.8 (*left*) plots the evolution in motion compensation during the first 200s of a test. For this figure, the algorithm started without any knowledge on the sensory system. Neither the orientation of the IMU nor sensor sensitivities were programmed. Therefore the plot starts with no compensation at all, then the axes are identified and sensitivity adjusted. Fig. 5.8 (*right*) shows the actual changes in gain for the gyroscope mainly corresponding to pan. The direct correlation between an axis of the camera motion device and a gyroscope is not necessary, the IMU can have any random orientation towards the CMD.

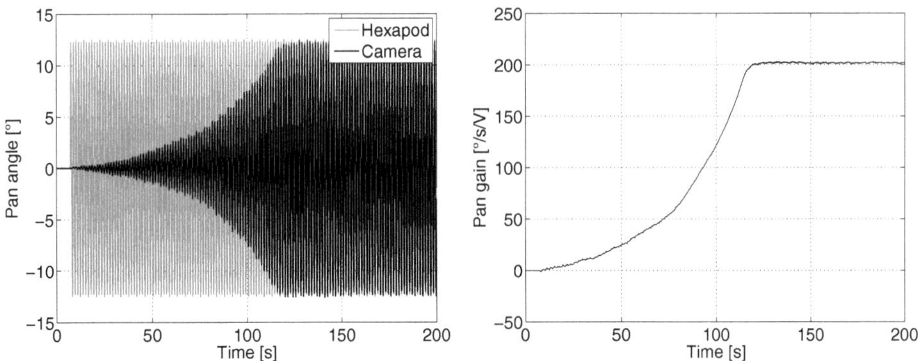

Fig. 5.8: Angular position of the hexapod and compensatory amplitude of the CMD (*left*); adaptive variation of the corresponding gyroscope gain (*right*).

With a final and constant gain reached after about two minutes, adaptation showed the gyroscope sensitivity to be within a 1.5% range from nominal values.

Since optic flow is too high for exact measurement in the beginning of adaptation with no prior knowledge, like in Fig. 5.8, a second test series shows the improvements when the sensor orientation is known and the system operates with nominal characteristics. Under these close-to-ideal conditions, the video subjectively seems very unstable. Image processing calculates an optic flow of more than 6 pixel per frame under laboratory conditions. Fig. 5.9 (*left*) shows the evolution of the absolute values of peaks in optic flow over time. In comparison to the nominal setup, flow could be decreased by more than 80%.

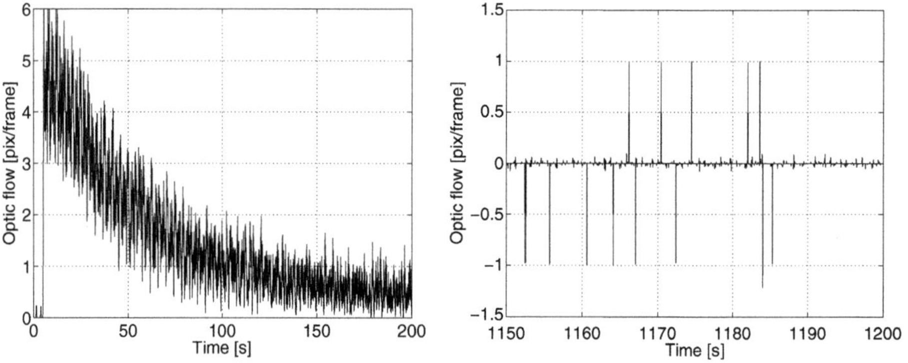

Fig. 5.9: Absolute values of peaks in optic flow for pan axis (*left*); optic flow projected onto tilt axis after 20 minutes of adaptation (*right*).

The best results could be achieved in tilt direction. Due to the direct coupling of the tilt frame to the gearbox without any pushrod in between, the kinematic structure has no backlash, which is crucial for a stable image. After 20 minutes of adaptation, image shifts of only one pixel per frame every few seconds, which corresponds to the resolution of image processing, could be detected in flow projected on the tilt axis.

5.2.4 Further Enhancements for Dedicated Applications

The presented algorithm is well suited for applications considered within the FOR-BIAS project. But in different environments, a slightly modified algorithm can perform better. This is especially the case in presence of broadband excitation. Then, the PT_1 approximation of the plant may be insufficient. Therefore, the structure of Fig. 5.3 is enhanced by a second, compensatory filter parallel to the adaptive path. It is designed with a standard tapped delay line FxLMS (TDL) structure, thus it receives the same input signals with different delays [142]. Due to the different delays, there is always one signal in the right phase to compensate for disturbances at every considered frequency.

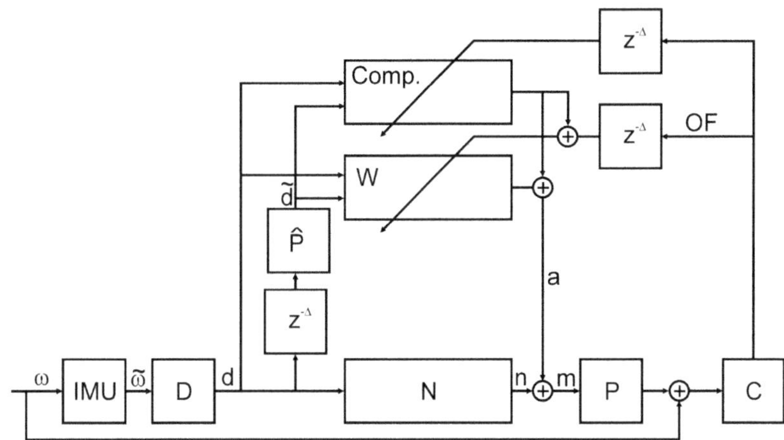

Fig. 5.10: Modified structure of the camera stabilization algorithm.

The basic idea is to use the structure of the previous sections for general camera stabilization and to have the second compensatory path for fine-tuning and, with a relatively high learning rate and leakage factor, for quick response to changing conditions. Therefore, the output of this path is used as a teaching signal of the basic adaptive path, for the latter can take over the low frequency changes from the tapped delay line LMS.

Fig. 5.11 shows the results of simulations with this setup. With an initial sensor gain reduced to 90%, the TDL immediately compensates for the deficit, but its output decreases with adaptation of the standard structure (*left*). At $t = 100s$ until $t = 200s$ a second source of excitation at a different frequency is added. Again the same, the TDL LMS is the first to respond and decreases afterwards.

This behavior is reflected in the residual optic flow, Fig. 5.11 (*right*). In comparison to the previous algorithm, the quick response of the TDL reduces the OF right from the beginning. But the main difference comes in the middle part in the presence of several frequencies. Due to the imperfect modeling, the standard algorithm compensates for a mean frequency, thus not completely stabilizing the camera. The TDL approximates the actual plant behavior better. A major drawback of tapped delay line LMS algorithms are sudden, shock-like changes in excitation since weights are redundant. Although a leakage factor for the weights is included, the TDL LMS produces a short peak in OF before it is adapted.

The enhancement presented in this section is suitable for applications where often varying excitations in a broadband range occur. For the FORBIAS applications, this is neglected since the first eigenfrequencies of the vehicles, especially in the car, are defined by the vehicle frames and dominant. In the Audi A8 used for the test drives, for example, little motion above 2Hz needs to be compensated for.

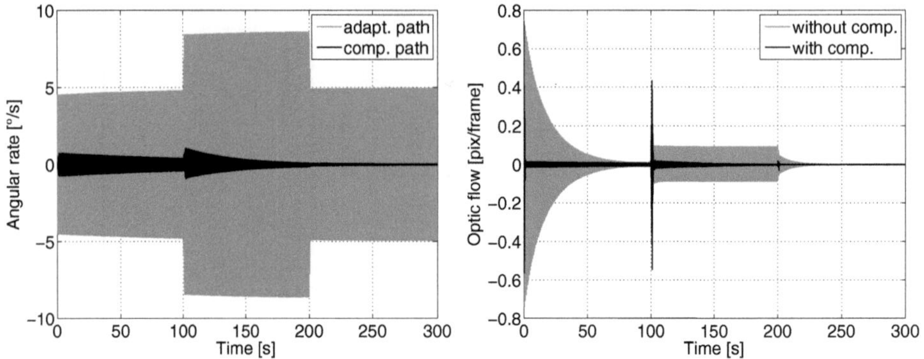

Fig. 5.11: Output of the two adaptive pathways (*left*) and residual optic flow (*right*).

5.3 Closed-loop biomimetic Gaze Stabilization

Until now, it was always assumed that an external source will provide the line of sight for the stabilized camera. This requires a second superordinate system like a second camera with a wider aperture angle in order to detect points of interest, or a special navigation system like in the helicopter application in Section 6.2. However, there are many applications in which it is desirable not to need an external reference. This can include assistance systems with only one (pivotable) camera, head mounted cameras for target tracking or navigation of autonomous robots, possibly in connection with self location and mapping (SLAM) algorithms. Another interesting application is a camera mounted on the end effector of a robot for visual guidance of the robot arm and increased positioning accuracy in assembling processes.

For these tasks, the stabilized camera needs to track the point of interest itself. The fact of closing the control loop brings two major challenges that need to be addressed. The first one is stability, since the tracking command from image processing arrives up to 100ms after an event occurred, which can destabilize the algorithm. The second problem, which is more important for the general topic of this work, is in the interference of visual tracking and adaptation in inertial stabilization, in neuroscientific terms between the opto-kinetic and the vestibulo-ocular reflex. If only the optic flow was used as a teaching/error command as described in the previous sections, the sensory gain would be erroneously decreased. During lowest frequency motion, the visual pursuit system dominantly takes over the task of image stabilization, leading to misadaptation of the sensory gains. But if the sensory/inertial stabilization is required after a while, for example due to a bump or shock, the misadjusted sensory gains cannot compensate for the movement. Therefore a different model with at least part of the teaching signal being different from optic flow is mandatory.

5.3.1 An adaptive model of the vestibulo-ocular system

The adaptive model developed here is based on the extensive prior work of GLASAUER and MARTI [49, 50]. When building the model, both researchers focused on studying the downbeat nystagmus [96, 95]. Due to the close relation between monocular camera stabilization and the biological system, which performs exactly the same task, the model is ideal for introducing adaptive VOR components [55]. The model is also based on knowledge of the neurophysiologic structure of the brain, as outlined in Section 2.2.

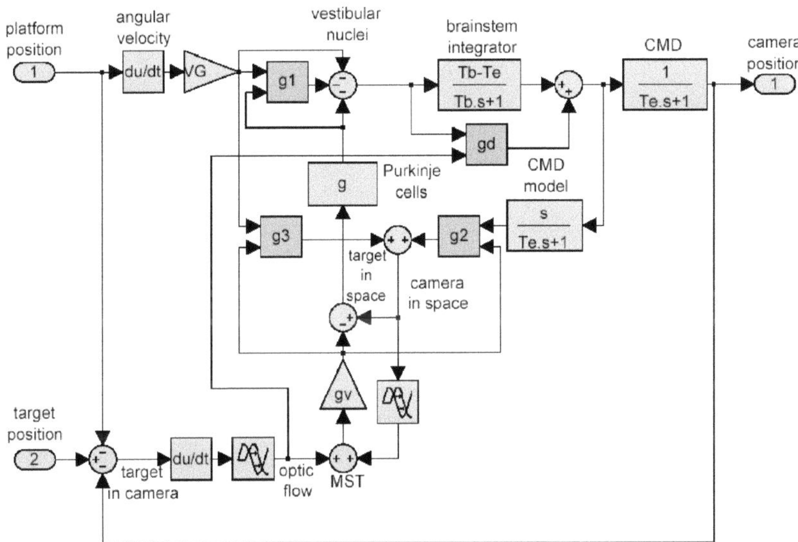

Fig. 5.12: One-dimensional biomimetic model for adaptive gaze stabilization with four sites of plasticity.

Input to the simulation model are platform (e.g. head position), with its derivative angular rates being measurable by gyroscopes, and visual target position in space. The latter is used to calculate the target position in the camera frame, with optic flow as the derivative. Again, the vestibular gain VG allows for artificially decreasing sensor gain to study adaptation. The upper pathway in Fig. 5.12 corresponds to the nominal compensation described earlier. In the central part, inertial sensor information and an estimate of the CMD motion are used to calculate the angular velocity of the camera in space. After passing through a delay, which gets the information in phase with visual information that arrives too late due to image processing and thus stabilizing the algorithm, an estimate of target motion in space is reconstructed with measurable data. MST indicates the location of this structure in the medial superior temporal area in the human brain.

The important parts are the identified sites of plasticity and the corresponding teaching signals. Like in the models before, a derivative gain gd is placed parallel to the integrator to compensate for the CMD dynamics. Optic flow is suitable as a teaching signal. The forward oriented gain $g1$ adjusts for VG deficits in the direct compensation path. Purkinje cell output showed to be a good adaptation signal and is physiologically plausible. For the gains $g3$ and $g2$ in the forward and recurrent paths contributing to the visual guidance command, the MST output achieves the desired adaptation.

The model as outline in Fig. 5.12 is intended to reflect the biological system as much as possible. Its one-dimensional structure can be expanded to the three dimensional case in the same way as the models in Section 5.2 with the corresponding Jacobians for inertial sensors and optic flow.

5.3.2 Simulation of Inertial Stabilization and Target Pursuit

To evaluate the performance of the model and to test convergence under all possible circumstances, simulations with a large variety of input parameters, including noisy signals were performed. In the beginning of each cycle, the visual input was interrupted (e.g., simulating motion in complete darkness), in order to determine the actual VOR gain. Then, the algorithms adapted for 200s, before the new VOR gain was determined. Fig. 5.13 shows the compensation during these VOR-only periods. The actual VOR gain increased for artificially decreased 67% up to 95%.

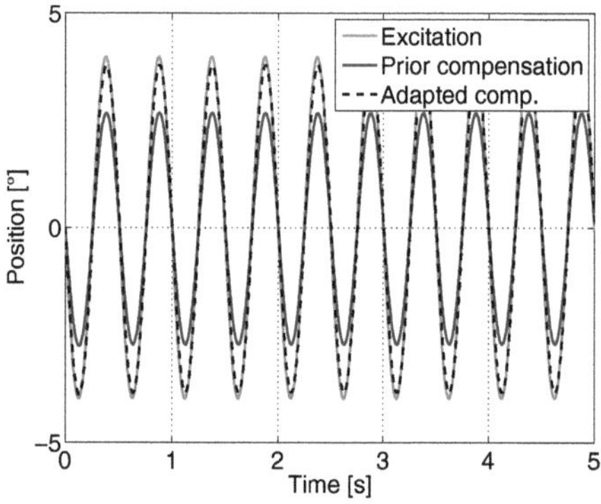

Fig. 5.13: Simulated VOR-only compensation prior to adaptation and after 200s of adaptation.

During the adaptation period, the four variable gains changed according to Fig. 5.14. The derivative path gain increases to a value corresponding to the time constant of the camera motion device. The two forward gains compensate for the vestibular/sensory gain. The recurrent gain only changes slightly. This site of plasticity has its main function when adjusting for deficits in the motor or visual path of the biological system.

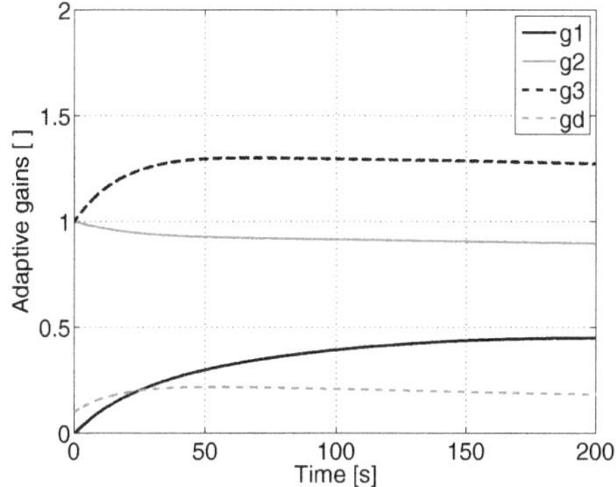

Fig. 5.14: Simulated VOR-only compensation.

The exact shape of the gain curves depends on the actual learning rates and their ratio between each other. The final result, however, is reproducible if the general restrictions of LMS algorithms are complied with.

5.3.3 Experimental Validation

For the experimental validation of the monocular camera stabilization, the same setup as shown in Fig. 5.6 and Fig. 5.7 was used with only slight modifications. The information sent from image processing to the gaze control now contains the position of the tracked target, extracted from the camera image.

Two series of tests show the performance of the model. First, the sensor sensitivity, equivalent to the vestibular gain, was artificially decreased to 60% and 70%. The adaptive structure of Fig. 5.12, duplicated for the pan and tilt axis, provided two-dimensional stabilization. Due to restrictions in the calculation of target rotation, the third axis was excluded from the test, but could be incorporated in the same manner.

Fig. 5.15 plots the evolution of all four gains during the first 120s of adaptation. For better comparison, the target the camera focused on, was kept still. The plots of gains $g1$, $g2$ and $g3$ outline the different adjustments for the different sensory deficits. It is obvious that a the higher sensitivity reduction to 60% produces higher changes in gains. The gain of the derivative path gd converges to the same level in both tests since it compensates for the dynamic behavior of the camera motion device, which stayed the same.

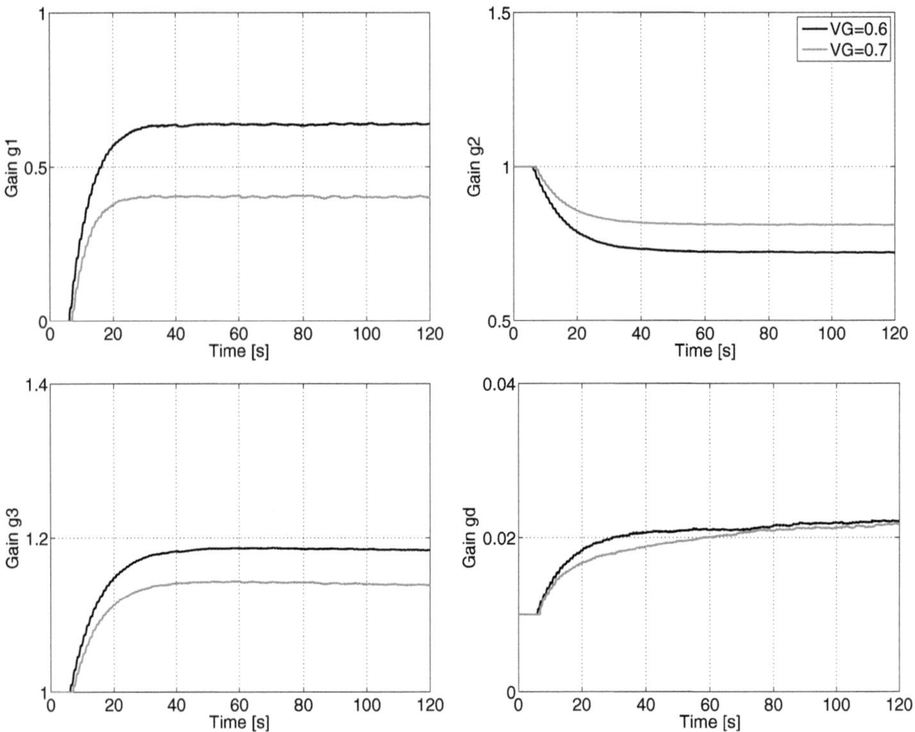

Fig. 5.15: Evolution of adaptive gains corresponding to pan axis for artificially decreased vestibular/sensory gain to 60% and 70%.

In the second series of tests the influence of target motion was studied. Fig. 5.16 compares adaptation of pan axis gains at a sensory gain setting of 60% for fixed and moving targets. Due to the hexapod trajectory, the camera platform was excited in exactly the same manner for both tests. The target itself was moved manually and randomly in the laboratory.

The general shape of the adaptation curves does not change for moving targets. Temporarily, there are slight deviations observable, especially for gain $g1$, but these short term effects can be reduced with a decreased learning rate and have no negative effect on stability of adaptation.

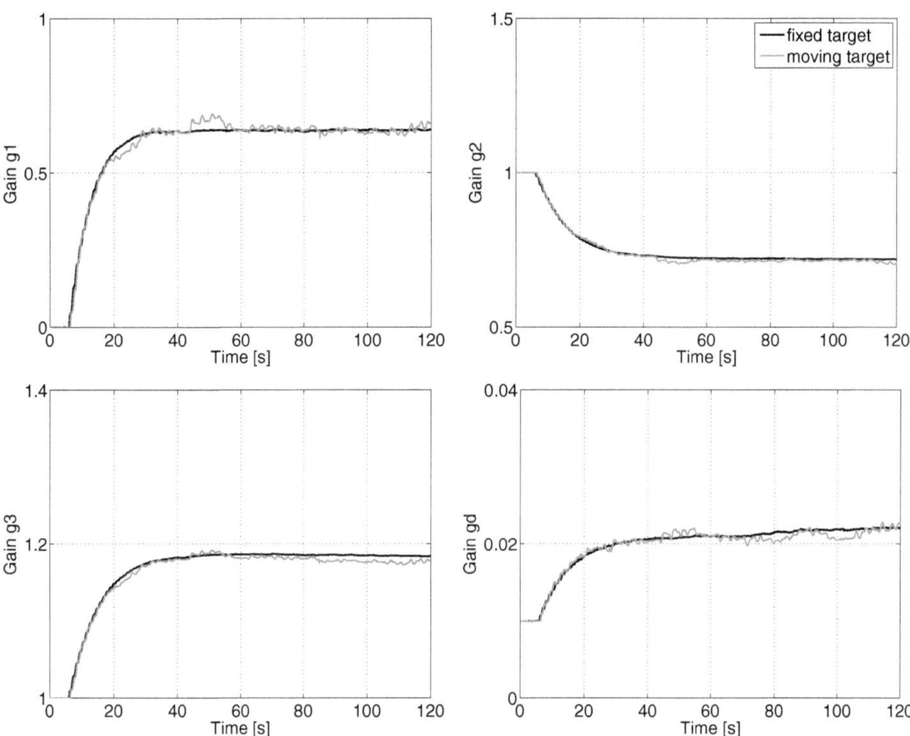

Fig. 5.16: Influence of target motion on gain adaptation for $VG = 0.6$.

6 Active Vision for Assistance Systems

6.1 An Inertially Stabilized Camera Platform for Driver Assistance Systems

6.1.1 Introduction

In the hard-fought market for upper class vehicles, car manufacturers have several ways to distinguish themselves and secure a competitive advantage. Together with the general effort to increase energy efficiency, the main fields of research are comfort and safety.

An efficient way to approach both topics is to strive for high-level driver assistance systems. Recently, *BMW* and *Daimler* introduced infrared camera-based night vision, which was the first time a video image of the environment had been presented to the driver. Suppliers like *Continental* developed vision-based lane detection and lane keeping based on low-resolution cameras. Until now, all approaches had focused on the use of wide-angle cameras for the perception of a large part of the environment.

A logical next step is the introduction of cameras with telephoto lens and a large focal length, in order to get a detailed view on specific points of interest, with the obvious need for active inertial stabilization as outlined in the previous chapters. Already a small bump in the road can induce vehicle motion large enough to make the image of a camera with an aperture angle of less than $8°$ useless for image processing. Such a pivotable camera would have many value-adding advantages:

- **Prevention of dangerous situations**
  When a driving situation is unclear and radar reflexions as well as the wide-angle image do not resolve the environment sufficiently, a detailed image of certain spots can provide life-saving information for a driver assistance system to warn the driver, or if necessary and allowed by law, even take over control to avoid a collision. This is especially suitable if the danger arises from a cyclist, pedestrian, or for example playing children close to the road.

- **Enhanced vehicle tracking**
  For high-level driver assistance systems reliable vehicle tracking is necessary. The considered image processing software uses abstract form models of cars [103, 104]. An additional camera with telephoto lens provides more reliability

W. Günthner: *Enhancing Cognitive Assistance Systems with Inertial Measurement Units*, Studies in Computational Intelligence (SCI) **105**, 101–129 (2008)
www.springerlink.com

for increased distances to an object. With the possibility of getting a high resolution image of the number plate, absolute identification of a vehicle allows for long-term tracking.

Fig. 6.1: Vehicle tracking and traffic sign recognition with telephoto camera lens.

- **Traffic sign recognition**
  The early detection of traffic signs like direction indications and speed limits favors a smooth vehicle navigation and control when this information is used in semi-autonomous driving. Lane changes and decelerations can be scheduled well in advance.

- **Vision based vehicle navigation**
  Current navigation systems rely on GPS signals, in car-fixed systems enhanced by wheel sensor and/or magnetic field information. During GPS outings, these secondary sources are the sole reference for navigation, but are suitable only as a short-term reference. Visual information can bridge a significantly longer time gap since it is not subject to "integration" like velocity signals. As a high-end reference, DARPA Grand Challenge vehicles navigate with cameras not only during GPS outages but also in an unknown environment [130].

In government-oriented rather than consumer applications, this system of vehicle tracking and number plate recognition can be used for automated traffic controls from onboard a police car. The patrol car just has to drive with the traffic flow: the system detects cars, checks number plates for comparison with a database and can additionally perform speed checks in an efficient way. A more driver-friendly application is the automated checking of road conditions. Highway maintenance can automatically detect, document and report defects in the road or markers using very few staff.

6.1.2 System Set-up

For the evaluation of the developed camera stabilization system, emphasis is on the suitability or transferability to series-production vehicles. To ensure operability under every-day driving conditions and a maximum of comfort for the vehicle occupants, some basic requirements must be met:

- **Range of operation**
  For exact motion-compensation, the inertial measurement unit, as well as the camera motion device, need to cover angular rates and velocities a car can experience. According to the participating industrial partner BMW, these are $1 rad/s$ and $1.6 rad/s^2$ under regular driving conditions. In terms of maximal amplitudes, the pivotable camera needs to cover $\pm 45°$ in pan as well as in tilt direction. Due to comparatively small roll angles, this third active axis is renounced.

- **Dynamic properties**
  The dynamic response of camera stabilization needs to meet at least the first eigenfrequency of the car. For the test vehicle, an AUDI A8, this is about 2Hz as shown in Fig. 4.6.

- **Object tracking**
  In addition to pure inertial stabilization, the system must be able to fixate specific objects in the environment. A continuous velocity-based tracking algorithm shall prevent saccades for purposes other than gaze shifts.

- **Gaze shifts**
  In order to keep track of multiple objects, the pivotable camera needs to perform gaze shifts at speeds of 500 °/s, comparable to the human oculomotor system. These saccades assure that a point of interest is in focus without loosing valuable reaction time for the driver assistance system's actions, as according to Section 6.1.1.

- **Mounting**
  The cameras and the IMU shall be mounted in a location that ensures a good overview of the environment. An ideal spot is between the windshield and the rearview mirror since it is high above the road and out of sight of the driver.

- **Communication**
  Until now, automobile manufacturers rely on the CAN bus protocol. Information exchange between the vehicle telemetry and the camera stabilization system shall use this interface.

- **Control algorithm**
  For a cost-efficient series application, the control algorithm of the camera stabilization shall be suitable for PowerPCs. This means avoiding computationally expensive mathematical operations.

As a result of these restrictions, a modular setup is ideal for enlarging the system according to car manufacturers' demands and available resources. The core of the developed camera stabilization is a dSPACE DS1103 PowerPC, Fig. 6.2, on which runs the adaptive algorithms described in Section 5.2 and enlarged in Section 6.1.4. It receives the inertial data from the μCube via a high-speed 1 Mbit/s CAN2.0A bus [56, 58]. In order to keep latency low and provide fast reaction to vehicle motion, the frame rate of the μCube is set to $1kHz$, acquiring all inertial sensor readings every millisecond. Due to the high baudrate of the CAN bus, transmission to the PowerPC takes only about $111ms$.

As a camera motion device, the pan-tilt unit developed by WAGNER and already used for the test on the Steward-Gough platform in Section 5.2.3 is suitable for the vehicle application. With angular velocities of more than 500 °/s and a bandwidth of more than 6Hz [41], it fulfills the technical requirements for the vehicle camera. Since it is too large for series applications due to its design envelope of more than 150x100mm, a new mirror motion device (MMD), which moves a mirror in front of the camera and keeps the camera still, is currently under development and testing [57, 60, 137].

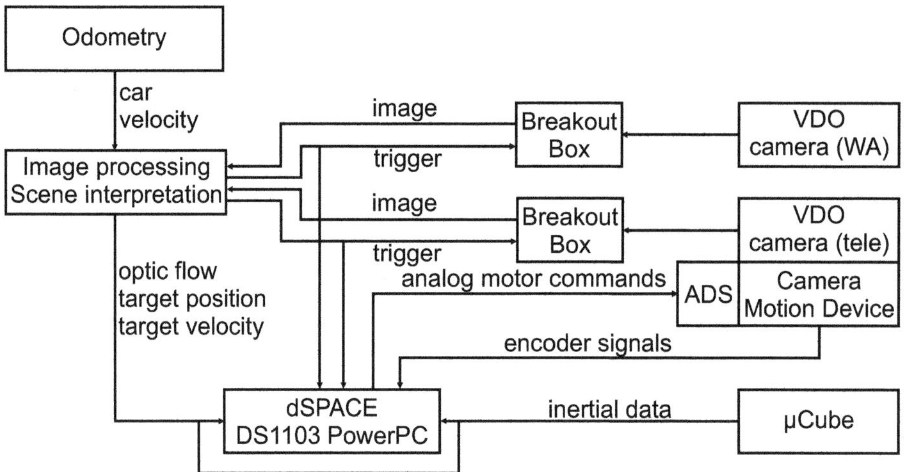

Fig. 6.2: Schematic diagram of the system setup of target detection and camera stabilization.

The visual setup consists of two identical Siemens VDO cameras with different lenses. For the wide-angle camera (WA), a focal length of $9mm$ is suitable to picture an overview impression of the environment. The pivoted camera has a $25mm$ telephoto lens for more detailed images. Both cameras feature a CMOS image sensor with an effective resolution of 750x400 pixels and a pixel size of $11\mu m$. The global shutter avoids artifacts that could harm image processing. The connection to the breakout boxes has an undisclosed proprietary protocol. From the breakout box to the master

PC of image processing, the video is transmitted in C-Link protocol. In addition to shutter time and other control parameters, the PC sends a trigger command to the cameras. This trigger is also received by the camera stabilization control in the dSPACE PowerPC in order to know exactly when an image was taken and how much time has passed. This is especially important for the prediction of the current line of sight (LOS) since the transmitted LOS, e.g. the target position, is already $50ms$ old when it arrives and is thus not precise any more.

The two video streams are captured in the image processing PC, that also received telemetry data and reconstructs a three dimensional model of the environment [104]. From this model, the position of the point of interest at the moment the images were taken is extracted in the camera platform coordinate system. In addition, the target velocity in space is estimated for smooth pursuit motion of the camera. As an adaptation signal for stabilization of the camera with telephoto lens, the optic flow of this video stream is also calculated in the same PC and sent to the central dSPACE control unit via CAN bus, together with information on the target. Since the update rate of visual information depends on a camera frame rate of about $25Hz$, this only generated a few CAN messages which are not time critical in terms of microseconds. So they are transmitted on the same CAN bus as inertial data, resulting in an overall bus load of less than 15%. Except for the hardware needed close to the cameras behind the windshield, all electronics are stowed in the trunk of the test vehicle, Fig. 6.3.

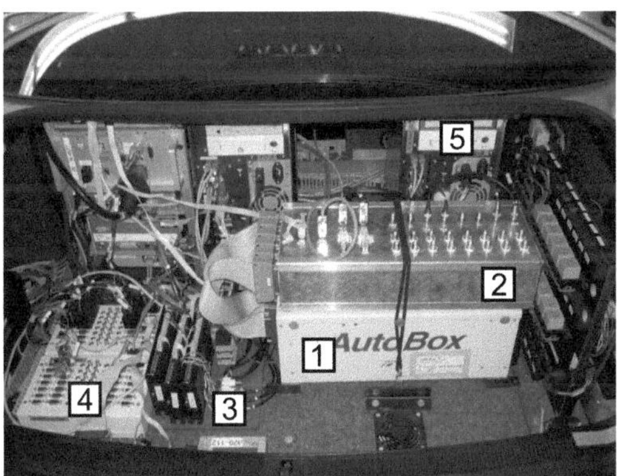

Fig. 6.3: Trunk of the test car (AUDI A8) with dSPACE real time control PPC (1), dSPACE breakout box (2), ADS (3), access to vehicle CAN bus (4) and image processing/scene interpretation PC (5).

For the protection and vibration isolation of the dSPACE DS1103 PowerPC, it was mounted in the AutoBox already installed by AUDI and supplied with the regulated 12V DC power from the onboard system. Together with the dSPACE breakout box, this part is most voluminous but allows for miniaturization in a series application. Currently, image processing and scene interpretation, e.g. the update of the environment model, is implemented on a standard PC (Fig. 6.3, 5). This allows for convenient integration in the Audi framework. In a future series application, the simple CPU, together with some peripheral electronics, will be sufficient.

Fig. 6.4 shows the front view of the vision system in the passenger compartment. On a stiff girder are mounted the inertial measurement unit (1), the car-fixed wide-angle camera (2) and the pivotable camera with telephoto lens in the pan-tilt camera motion device (3). Additionally, the screen for controlling the image processing PC is visible (4).

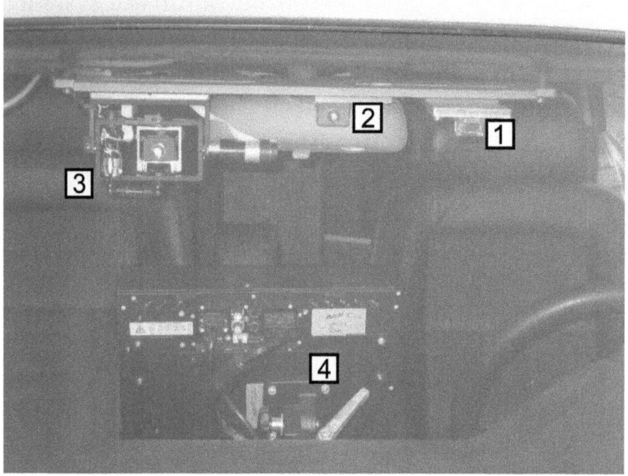

Fig. 6.4: Front view with μCube (1), wide-angle camera (2), pivotable camera with tele-photo lens (3) and control monitor for image processing (4).

6.1.3 Model-based Target Detection

Scene interpretation with the wide-angle camera is based on the 4D-approach initially proposed by DICKMANNS [28, 29]. Instead of the computationally extremely expensive extraction of 3D objects from a 2D image, it takes advantage of the time continuity of objects in the real world. Object attributes and characteristics are propagated with a dynamic state model and projected into the following image. Due to this expectation, the region of interest in the new image is significantly reduced, leading to lower computational demands. With the feedback estimation error, a recursive algorithm gradually updates model assumptions [104]. The algorithm used in the FORBIAS project and developed by NEUMAIER [103] reliably detects and tracks lane markers. By differentiating solid and broken lines, orientation on the road is possible. In connection with the test vehicle's radar sensors, other cars are detected and tracked with 3D form models. This visual tracking of cars has a better positioning accuracy in directions normal to the line of sight and can also account for still-standing cars which are usually suppressed with radar sensors.

6.1.4 Fusion of Visual and Inertial Information

The control for adaptive camera stabilization presented here is based on the algorithm developed in Section 5.2 and summarized in Fig. 5.3, reduced to the two degrees of freedom pan and tilt of the camera motion device. The major difference is that now the camera is no longer stabilized around neutral position but a specific line of sight is externally imposed. Additionally, the camera smoothly tracks points of interest (POI), velocity-based.

Therefore the coordinates of the POI are determined from the environment model, expressed in the camera platform system, which coincides with the coordinate system S of the inertial sensor and sent to the camera control on the dSPACE PowerPC. In order to calculate the required camera angles, the vector from the camera to the POI is normalized:

$$
{}_S\mathbf{r}_{POI,norm} = \begin{pmatrix} {}_S r_{POI,xn} \\ {}_S r_{POI,yn} \\ {}_S r_{POI,zn} \end{pmatrix} = \frac{1}{\sqrt{{}_S r_{POI,x}^2 + {}_S r_{POI,y}^2 + {}_S r_{POI,z}^2}} \begin{pmatrix} {}_S r_{POI,x} \\ {}_S r_{POI,y} \\ {}_S r_{POI,z} \end{pmatrix} \quad (6.1)
$$

Note that for consistency with the other involved systems in the vehicle, the x-axis points left with respect to the forward direction of the car, the y-axis upwards and the z-axis forward. Since the camera points exactly in the z-direction of its C-system, the normalized vector to the POI is the third column of the transformation matrix (Eq. 5.12) between the camera system and the sensor system, in which it is expressed:

$$
{}_S\mathbf{r}_{POI,norm} = \begin{pmatrix} \sin\phi \\ -\sin\psi\sin\phi \\ \cos\psi\cos\phi \end{pmatrix} \quad (6.2)
$$

From this equation, the desired pan angle ϕ_{des}

$$
\phi_{des} = \operatorname{asin}({}_S r_{POI,xn}) \qquad \phi_{des} \in [-\pi/2; \pi/2] \quad (6.3)
$$

and tilt angle ψ_{des}

$$
\psi_{des} = \operatorname{asin}\left(-\frac{{}_S r_{POI,yn}}{\cos\phi_{des}}\right) \qquad \psi_{des} \in [-\pi/2; \pi/2] \quad (6.4)
$$

are calculated with the mechanical restriction of $\pm 45°$ in both degrees of freedom of the CMD. However, the information on the POI position and velocity arrives with a non-deterministic delay due to the latency of the image processing and scene interpretation. In a representative test drive, the delay Δt varied from $75ms$ to about $87mm$, with an average of $82ms$ and some samples well outside the limits of Fig. 6.5. Therefore a variable prediction for delay compensation can significantly improve the performance. At a speed of, for example, $180km/h$, a car travels $4m$ during this visual delay, resulting in a significant shift in the desired line of sight.

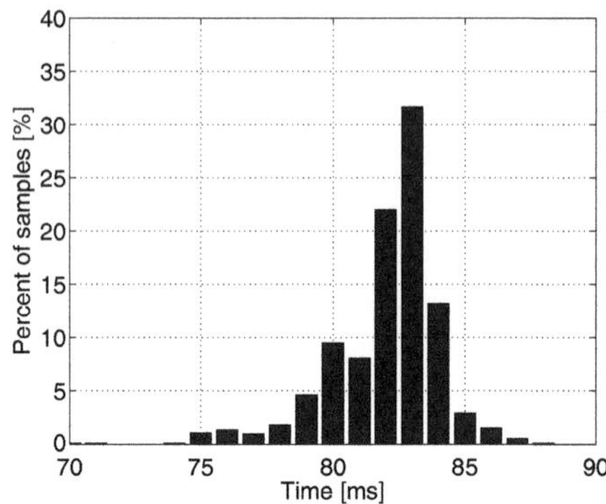

Fig. 6.5: Latency distribution of the visual path.

Since the camera stabilization control receives a copy of the camera triggers (Fig. 6.2), the delay until the LOS arrives via CAN bus from image processing is measurable. In a first order prediction, the coordinates of the POI are calculated with the difference of velocities in space of the POI, as derived from the environment model, and of the vehicle $_{S'}\mathbf{v}_C$. Note that a traffic sign or a lane marker has velocity zero.

$$_{S'}\mathbf{r}_{POI} = {_{S'}}\mathbf{r}_{POI,M} + \left(_{S'}\mathbf{v}_{POI,M} - _{S'}\mathbf{v}_C\right) \cdot \Delta t \tag{6.5}$$

In addition, the vehicle itself may have rotated during the delay time. Since the rotation angles between the sensor system at the moment the camera images are taken, S', and the moment the CAN message arrives, S, are small, a simple cardan transformation matrix, $\mathbf{A}_{SS'}$, with integrated angular rates, provides sufficient accuracy.

$$_{S}\mathbf{r}_{POI} = \mathbf{A}_{SS'}(\alpha = {_{S'}}\omega_x \cdot \Delta t, \beta = {_{S'}}\omega_y \cdot \Delta t)_{S'}\mathbf{r}_{POI} \tag{6.6}$$

In order to guarantee a minimum of saccades - that is, sudden gaze shifts that reduce the quality of the video stream - the tracking of points of interest is based on velocity and not position. With the relative velocity of POI and vehicle $_C\mathbf{v}_{POI,rel}$, expressed in the inertially stabilized camera frame, the required pursuit command is generated:

$$_C\boldsymbol{\omega}_{POI} = \frac{_C\mathbf{r}_{POI} \times {_C}\mathbf{v}_{POI,rel}}{|_C\mathbf{r}_{POI}|^2} \tag{6.7}$$

For Eq. 6.7, the vector or relative velocity is also transformed to the current sensor system S:

$$_S\mathbf{v}_{POI,rel} = \mathbf{A}_{SS'}(\alpha = {}_{S'}\omega_x \cdot \Delta t, \beta = {}_{S'}\omega_y \cdot \Delta t)_{S'}\mathbf{v}_{POI,rel} \tag{6.8}$$

The transformation from this sensory system to the camera system is performed with the inverse of Eq. 5.12 and the current CMD angles:

$$_C\mathbf{r}_{POI} = \mathbf{A}_{CS}\ {}_S\mathbf{r}_{POI}\ , \quad {}_C\mathbf{v}_{POI,rel} = \mathbf{A}_{CS}\ {}_S\mathbf{v}_{POI,rel} \tag{6.9}$$

Finally, the pursuit command is calculated with the projection on the degrees of freedom of the camera motion device, e.g. by multiplication with the corresponding Jacobian:

$$\begin{pmatrix} \dot{\psi}_{POI,pursuit} \\ \dot{\phi}_{POI,pursuit} \\ \dot{\gamma}_{POI,pursuit} \end{pmatrix} = \frac{1}{\cos\phi} \begin{pmatrix} \cos\gamma & -\sin\gamma & 0 \\ \sin\gamma\cos\phi & \cos\gamma\cos\phi & 0 \\ -\sin\phi\cos\gamma & \sin\gamma\sin\phi & 1 \end{pmatrix} {}_C\boldsymbol{\omega}_{POI} \tag{6.10}$$

The visually obtained information on the line of sight and the pursuit command are incorporated in the "integrator" block of Fig. 5.4. Apart from this, exactly the same stabilization algorithm as in the previous chapter is implemented in the vehicle system. Fig. 6.6 shows the enlargement to four inputs to the integrator for the tilt axis. The visual pursuit command $\dot{\psi}_{POI,pursuit}$ is directly added to the compensation command for pure inertial stabilization, $\dot{\psi}_{inertial}$. In accordance with Eq. 5.15, where the integrator is slightly leaky, the top pathway of Fig. 6.6 feeds the difference between the desired and the actual CMD angle to the integrator, causing a constant tendency towards the required angle and thus avoiding drift.

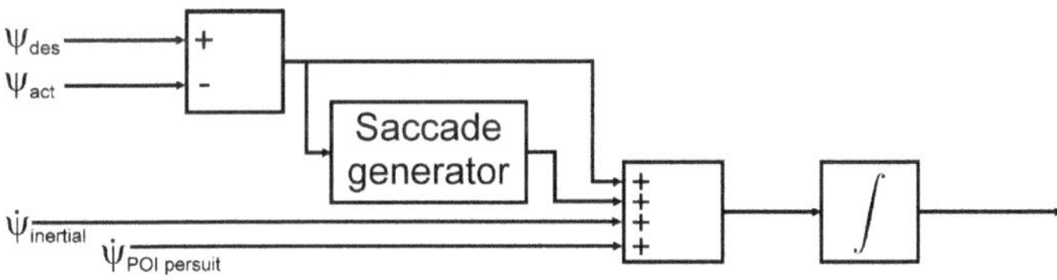

Fig. 6.6: Integrator block.

The saccade generator is responsible for quick gaze shifts. Its output is set to $\pm 300°/s$ if the input difference is more than the threshold of $\mp 1.5°$ and held until zero input is reached. This generates a fast saccade when the camera is more than $1.5°$ off the desired line of sight. In case of low quality in the velocity estimation for pursuit due to difficult visual conditions, the camera is rotated back to point at the target. In addition, intended gaze shifts are executed here since a new LOS usually exceeds the threshold for a saccade. The exact value of the threshold is a tradeoff between pointing accuracy of the camera and a stable image since a low value leads to frequent saccades which degrade the perception of the video.

6.1.5 Testing Scenarios

The test drives were designed to prove the operability of the complete camera stabilization system under real-world conditions according to the needs of future series
applications. Restrictions in the choice of the points of interest arose from the current status of implementation of object models in the image processing software;
only lane markers and other cars could be reliably tracked. Four representative scenarios cover the whole range of possible applications and allow for evaluation under
a wide variety of circumstances.

In a first series, no point of interest was defined but the camera with telephoto lens
was inertially stabilized. Without the interference of visual tracking and pursuit
commands, the quality of the inertial measurement unit and the camera motion
device could be assessed.

The second tests focused on the interaction between scene interpretation, pursuit
commands and inertial camera stabilization. Interlane marking about $40m$ ahead
was detected and tracked (*mode A*). By keeping a constant distance to this point of
interest, curvature of the road had the main influence, apart from stabilization.

As a variation on the previous tests, specific lane markers about $60m$ ahead of the
test vehicle were detected and tracked until they reached a minimum distance of
$20m$ from the car (*mode B*). Then the next marker, again $60m$ ahead, was chosen.
This series demonstrated the tracking of static objects on the road and can be seen
as a model for future tacking of traffic signs.

As the last scenario, other cars were tracked. This was the most complex task since
it involved detection of vehicles by radar sensors and image processing, accurate
tracking of the corresponding form models in the scene model derived from the
wide angle images, estimation of the other cars' speed, tracking of these random
movements and additional inertial camera stabilization.

To evaluate image processing under different visual conditions, the tests were performed with daylight, during dusk and by night.

The robustness was demonstrated by performing the tests on multi-lane freeways
at a speed of up to $180km/h$, as well as on small rural highways at much lower
velocity.

6.1.6 Testing in Road Trials

Together with AUDI, which was one of the project's industrial partners, the driving
tests in the vicinity of Ingolstadt proved the complete operability of inertial camera
stabilization and visual tracking.

Fig. 6.7 shows the system in operation on a freeway. For immediate quality control while driving, both videos of the cameras with telephoto lens and with wide-angle lens are displayed on the screen between the driver and the passenger seat. Apart from saving test data, no operator interaction was necessary during the test cycles. Due to data quantities of about $1GB$ per minute of driving, individual tests were limited to $120s$ each. For that reason, the inertial stabilization system was pre-adapted with the results from the test on the Steward-Gough platform, Section 5.2.3.

Fig. 6.7: Vehicle tracking and traffic sign recognition with telephoto camera lens.

The most significant data plots concern the tests tracking the lane markers since they combine predictable tracking motion with stochastic interference upon the vehicle's self-motion. In Fig. 6.8 data of the tests tracking one specific lane marker (*mode A*) are plotted.

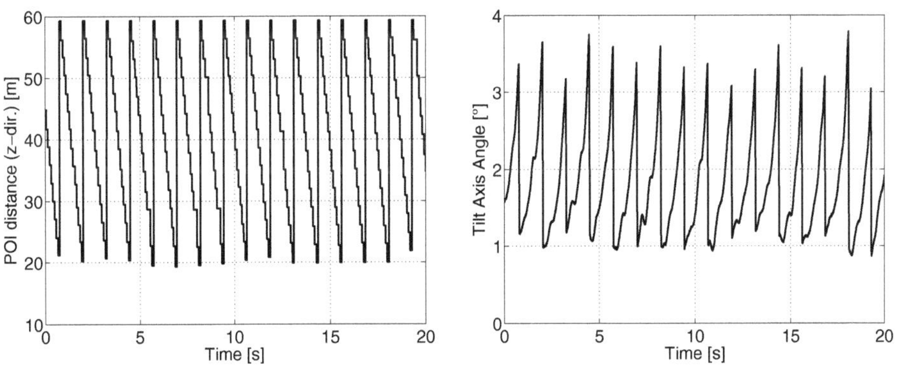

Fig. 6.8: Tracking of individual lane markers, distance to the car (*left*) and resulting tilt axis angle (*right*).

The left part shows the distance to the point of interest, the marker, in z-direction in the camera base system. Without loosing the POI, image processing determines a smooth and repeatable trajectory. Saccades are only initiated when a new marker is chosen. This is especially important to ensure unintended gaze shifts, which reduce video quality, are avoided. The right plot shows the resulting camera angle (tilt) of the CMD. For geometric reasons it is plausible that the angle increases faster when the marker is close to the car. In every cycle the additional stabilization commands are visible as deviations from the smooth angular trajectory.

Effects of inertial stabilization are more obvious in the test where lane markers at a constant distance from the car are tracked (*mode B*). Fig. 6.9 shows data of a representative 60s interval.

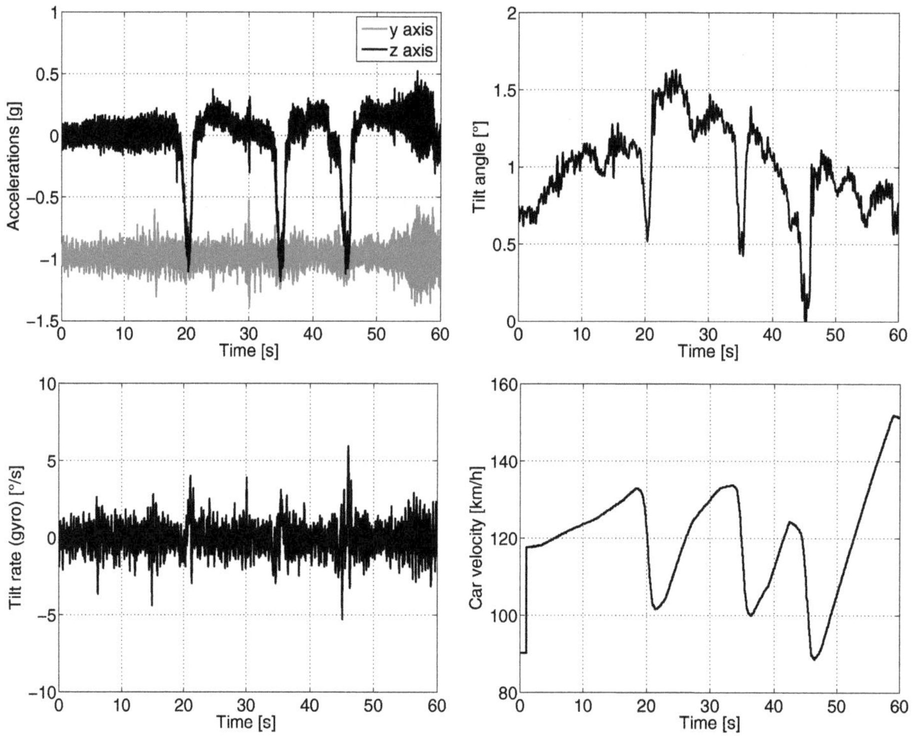

Fig. 6.9: Measurement data during test drives on a German freeway.

Again effects can be best shown for the tilt axis. In order to provoke an inertial stabilization reaction, the driver of the test vehicle accelerated and then braked sharply. This becomes obvious in Fig. 6.9 (*top, left*) where accelerations of about $-1g$ are measured in car longitudinal direction at $t = 20s$, $t = 35s$ and $t = 45s$. During these maneuvers the vehicle velocity was reduced by $30km/h$. As a result of braking, the front wheels deflected, leading to a vehicle rotation in tilt direction, measured by the corresponding gyroscope (*bottom, left*). The compensational motion of the CMD is visible in Fig. 6.9 (*top, right*) where a reduction in the tilt angle, thus an upward rotation, of about 1° is observable during the braking. The general level of the tilt angle starting at 0.5°, going up to 1.5° and then going down again, is the result of the curvature of the road. In the middle of the test sequence, an increasing descent leads to a higher camera angle.

Another remarkable point can be seen in the acceleration plot. Close to the end of this testing sequence, from $t = 55s$ on, the level of vibrations suddenly increases. This is the result of passing with the left tires onto the shoulder of the freeway, leaving the actual lane.

6.1.7 Discussion of Results

The driving tests showed that the specifications of the system, provided by the project's industrial partners, were far-reaching enough to cover all standard driving situations on highways, as well as on freeways. Since measured angular rates rarely exceeded 40°/s, gyroscopes with a low measurement range and a high resolution are clearly favorable. Therefore the μCube with ADXRS150 sensors is best suited, leaving a measurement margin for more extreme driving situations and resolving angular rates at 0.1°/s intervals.

Another important result is the determination of the actual frequency spectrum of motion and vibration occurring at the location of the camera. A representative Fast Fourier Transform has already been plotted in Fig. 4.6 of Section 4.2. It shows main disturbances peaking at about $2Hz$, which is sufficiently lower than the corner frequency of the camera motion device being about $4 - 6Hz$ [135]. With the currently used CMD, the higher frequencies around $18Hz$ and above $40Hz$ cannot be actively compensated. In a series application, a less rigid mounting of the inertial measurement unit and the camera can passively reduce the effects of those vibrations. But although they were measured in the conducted test drives, no influence on the camera videos could be observed.

The scene-interpretation algorithm reliably detected all different types of lane markers and thus properly identified the position of the test vehicle with respect to the road. Especially remarkable is the correct velocity estimation of the POI since errors in this data would lead to disturbing saccades. Fig. 6.10 shows the results of vehicle tracking experiments. After vehicle recognition with radar sensors, other cars were visually tracked in the wide-angle image. The two top images were taken by the car-fixed wide camera, a white cross marks the currently tracked vehicle.

The images in the second row are the corresponding views of the camera with telephoto lens. Especially the first picture shows how this pivotable camera contributes to the enlargement of the field of view. Due to the sharp, unblurred image in connection with a high focal length, the number plate is easily legible with image processing software.

Fig. 6.10: Vehicle detection in wide angle images (*top*) and corresponding images of the stabilized pivotable camera with telephoto lens (*bottom*).

The advantage of inertial stabilization becomes more obvious with longer exposure times of the CCD chip due to low illumination. Fig. 6.11 gives a detailed view on a traffic sign under night driving conditions.

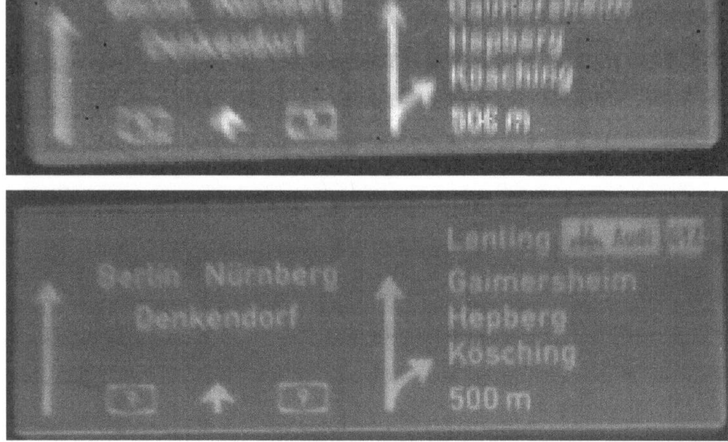

Fig. 6.11: View of a traffic sign at night with a car-fixed wide-angle camera (*top*) and the inertially stabilized camera with telephoto lens (*bottom*).

The lower resolution of the top image is mainly due to the larger field of view using the wide-angle camera; however significant image blurring is obvious and automatic sign recognition could hardly extract information. The lower image, taken with equivalent camera settings, is sharp and unblurred. This is even more remarkable since the small aperture angle of the lens makes it more critical. Visual navigation algorithms can receive all information, the number of the freeway, the directions and exits from this image.

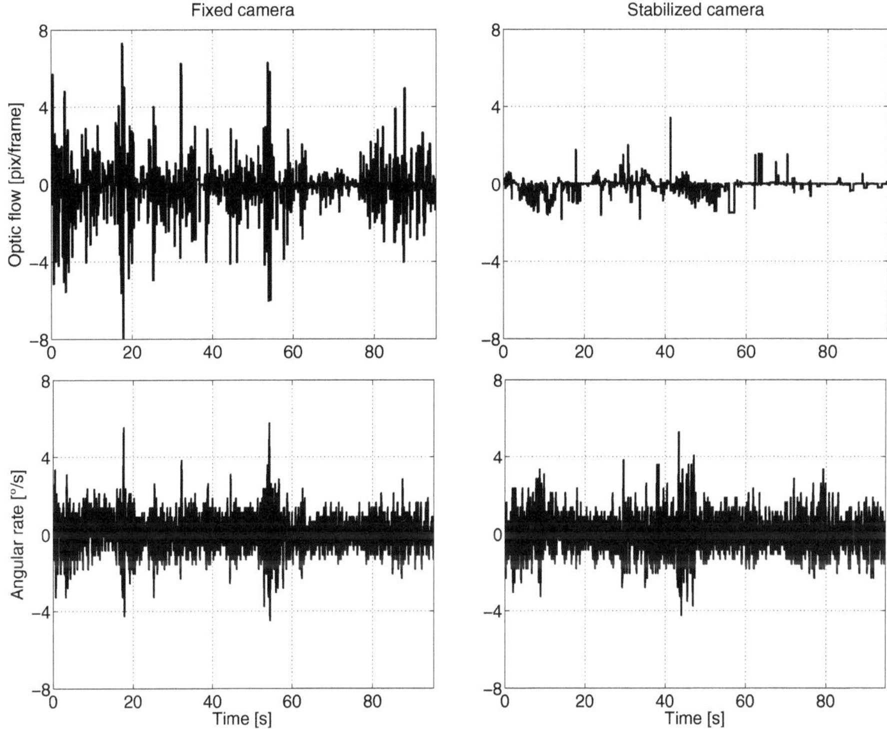

Fig. 6.12: Optic flow and angular rates during test drives with a car-fixed (*left*) and inertially stabilized (*right*) camera.

After these subjective impressions the quality of camera stabilization is quantified in Fig. 6.13. Two similar test cycles are compared, one with the regular stabilization of the tele camera (*right*) and one without stabilization, e.g. with the actuators of the camera motion device turned off (*left*). In both cases vehicle motion was induced by braking sharply. The total reduction of optic flow is obvious in the stabilized case, but more important is the decorrelation of optic flow and pitch rate. Whereas in the unstabilized case elevated angular rates at $t = 18s$ and $t = 55s$ lead to corresponding optic flow in the upper graph, car motion at $t = 45s$ in the stabilized case have no influence of image motion. Main optic flow in the latter case, negative during the first $20s$, then mainly positive for $20s$ and then negative again, has low frequency and is the result of overtaking cars, which also generate motion in the image.

This correlation becomes more obvious when the data from all four diagrams is combined into one, eliminating time. In Fig. 6.13 optic flow is plotted as a function of angular rates. For both cases a regression line is fitted to the data points using a standard least mean square algorithm. Note that a high number of data points close to zero optic flow leads to a lower slope of the line for the fixed camera than one would expect when looking at the diagram.

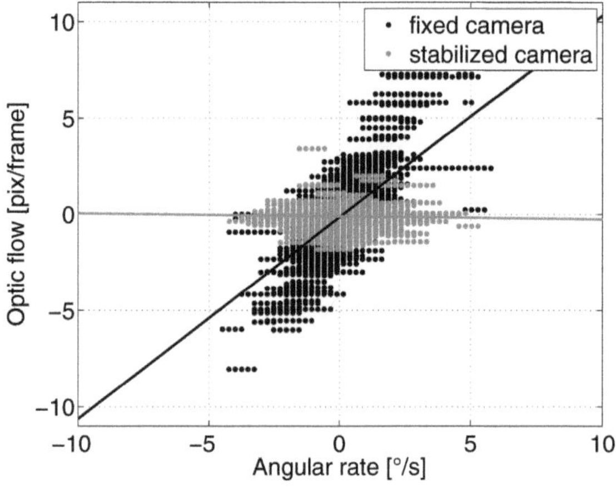

Fig. 6.13: Correlation of optic flow and pitch rates for a car-fixed and an inertially stabilized camera with telephoto lens.

The slope of the regression line characterizes the dependency of both variables. In the non-stabilized case there is a clear dependency, higher angular rates lead to higher optic flow. For the stabilized camera the slope is close to zero, signifying that flow is independent of the vehicle's self-motion, which in return means that the camera has perfect inertial stabilization.

6.2 Pilot Assistance for Instrument-aided Approach Scenarios

6.2.1 Introduction

The second industrial application of the stabilized camera platform was implemented in cooperation with EUROCOPTER. Current research on pilot and operator assistance for helicopters focuses on flying under harsh environmental conditions and in difficult surroundings.

Within the research cooperation PILAS, funded by the German Aerospace Research Programme III, EUROCOPTER is developing a pilot assistance system for *Helicopter Emergency Medical Service* missions under all weather conditions [75]. The basis is a high-level terrain map including obstacles. The interface to the pilot is a large screen in the cockpit showing the flight path as a semi-transparent tunnel in a computer-animated environment in real-time. However, the system does not use active vision for the detection of temporary obstacles. The camera platform developed by FOR-BIAS can significantly contribute to further increasing flight safety and reducing the pilot's workload under challenging conditions. In any case, it needs to provide a sensory advantage over existing systems or the pilot himself, such as by showcasing a high resolution or sensitivity to light not visible to the human eye.

- A pivotable camera with a high focal length can detect temporary obstacles in the proposed flight tunnel for the Pilas system. At high altitude, a radar-based system might be more suitable for this task; however, this does not work for static obstacles on the ground in the vicinity of the landing place.

- In a mountainous or otherwise harsh environment, a camera with telephoto lens can give an early visual impression of a point of interest to the pilot. This can be a landing place or victims during a rescue mission. Depending on the location of the camera in the helicopter, an image of the immediate environment otherwise not visible to the pilot can help during stationary hovering.

Fig. 6.14: Pilot support in a mountainous environment.

- For landing in an urban environment, such as a rescue mission involving an accident, a visual image can be used to generate a dynamic map of the site, supporting the pilot in his final approach.

- When equipped with a camera sensitive to infrared light, the pilot can be supported during the approach, showing, for example, people on the landing place.

Apart from pilot assistance, similar systems are already used for surveillance tasks. Federal Border Patrol detects people illegally crossing border lines with infrared sensitive cameras. Traffic violators can be detected from police helicopters and identified with a zoom on the number plate. For all of these current applications, high-precision and extremely expensive hardware guarantees accuracy.

The goal of this section is to show that the newly developed inertially stabilized camera platform can solve the tasks with low-cost hardware, taking advantage of the adaptive algorithms described in the previous chapters. Conducted test flights at the EUROCOPTER airport in Donauwörth/Germany prove its operability in final approach scenarios.

6.2.2 Helicopter System Set-up

For obvious reasons, the most important aspect in the design of the camera stabilization system for the helicopter application is flight safety. Under all circumstances, disturbing feedback to the helicopter flight control and instruments must be avoided. The second critical point concerns incidents relating to the hardware itself. EURO-COPTER requires all components to be in housings solid enough to contain every piece, even if it loosens completely.

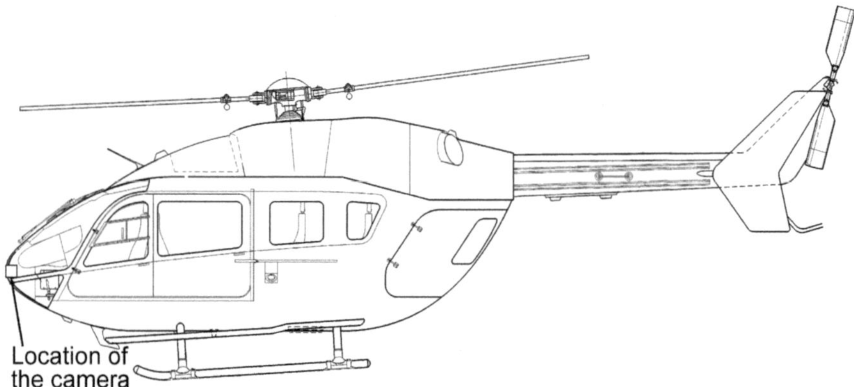

Fig. 6.15: Side view of the test helicopter, indicating the location of the FORBIAS camera platform.

These basic ideas lead to a separation of the hardware in two locations, similar to the test drives. The pivotable camera and the inertial measurement unit are mounted in the nose of the helicopter, Fig. 6.15, whereas all other electronics are in a 19" rack in the back, Fig. 6.16. The camera location in the front has the advantage of capturing an image similar to the view to which the pilot is accustomed. With a vertical view, the camera being mounted under the helicopter and facing down, the pilot could get confused in critical situations. In addition, a forward camera is most suitable for final approach scenarios, which are the basis of the flight tests conducted within this project.

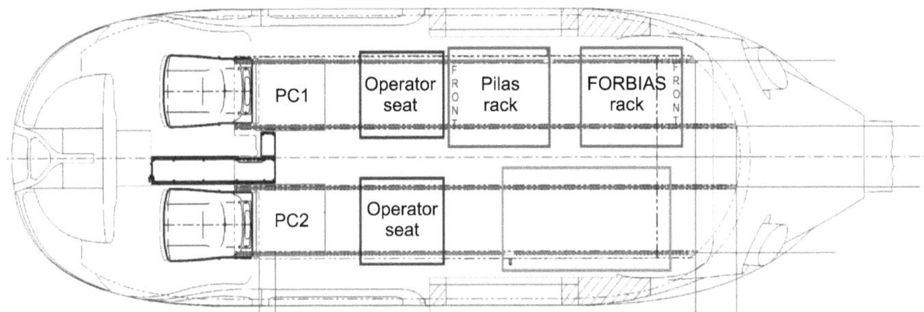

Fig. 6.16: Layout and setup of the test equipment inside the helicopter.

The front components are mounted in a mostly aluminum housing, bolted to the main frame of the helicopter. Inside the box, Fig. 6.17 (*left*), the inertial measurement unit and the camera motion device share a common ground plate, suspended with elastomere brackets. These passive vibration isolation measures reduce the effects of the high eigenfrequencies of the helicopter.

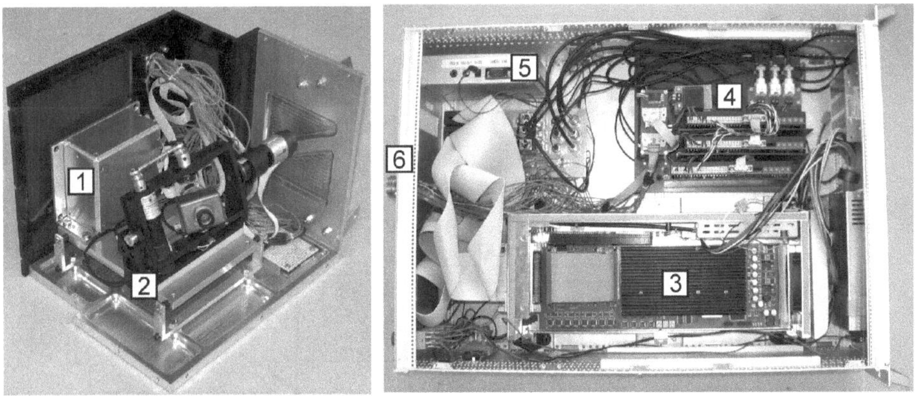

Fig. 6.17: Inertial measurement unit (1) and camera motion device (2) in a half-opened housing (*left*); top view of the 4 RU 19" rack drawer with dSPACE PowerPC (3), ADS (4), VDO breakout box (5) and single aerospace connector (6) (*right*).

Due to the mounting of the box to the outer surface of the helicopter, it is exposed
to all weather conditions. Special attention was paid to thorough waterproofing. For
fast installation, one single aerospace-approved connector provides the interface to
the onboard wiring of the helicopter.

As mentioned above, the control unit is placed in a rack in the rear part of the
helicopter. The most critical electronic components, the dSPACE PowerPC and the
ADSs, have passive vibration isolation. In addition, the dSPACE circuit board was
itself given an extra suspension. The breakout box for the Siemens VDO camera, as
well as all necessary power sources for the components, are placed inside the rack
drawer. A special high-speed ventilator keeps the temperature inside the rack within
the certified limits.

The wiring between the two locations of FORBIAS equipment is helicopter-fixed in
the cable duct. For the high-speed CAN bus connection of the inertial measurement
unit, a twisted cable pair is used. Power is supplied by the $28V$ board system and
has an individual buffer for the FORBIAS electronics.

Fig. 6.18 shows the system setup in terms of wiring and information flow. The
inertial measurement unit, the camera and the CMD use the same protocol as for
the test drives. The connection between the camera and its breakout box has a
proprietary and undisclosed Siemens VDO specification. The camera is controlled
by a EUROCOPTER PC via RS232, the video stream delivered back via C-Link. Image
processing, e.g. the calculation of the optic flow for the adaptive gaze stabilization
algorithm, is computed on the ECD Linux PC and sent to the control unit via CAN
bus.

Before the test flights started, the complete system was extensively tested by the
EUROCOPTER flight inspector.

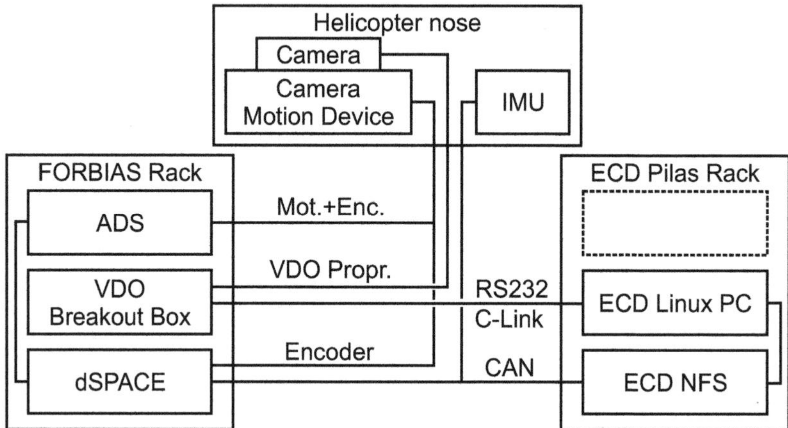

Fig. 6.18: Wiring scheme between the two FORBIAS components and the interface to
PILAS.

The main difference to the system as it is implemented for the automotive tests is the way how the point of interest, or the desired line of sight, is determined. This will be detailed in the following section.

Fig. 6.19 shows the test helicopter with the camera system installed for test flights.

Fig. 6.19: Front view of the test helicopter with installed camera system.

6.2.3 Navigation-based target detection

In contrast to the application with driver assistance systems, the point of interest is not determined from a camera with wide-angle lens but is calculated from information of the helicopter navigation system. The board computer has accurate knowledge on the absolute position of the aircraft in the World Geodetic System (WGS84), as well as on its orientation in space.

Since aeronautics and especially instrument-flight oriented navigation is very coordinate oriented, the point of interest, especially when it comes to landing places, is also known in world coordinates. For the test flights within this project, one point on the runway of the EUROCOPTER airfield in Donauwörth is chosen as the target for the stabilized camera. For rescue missions this could be, for example, an intersection or a building. For special cases where the coordinates of the POI cannot be determined in advance, the camera image itself in connection with the helicopter position can be used to calculate the POI, once it can be manually found with the image.

From these two positions and the helicopter orientation, the required camera angles of the desired line of sight are directly computed in the EUROCOPTER navigation system and transmitted to the FORBIAS control via CAN bus (Fig. 6.18).

Another major difference to the previously described tests is the way target pursuit is realized. Due to restrictions, it was not possible to extract the helicopter speed from the board navigation system and send it to the camera stabilization control. Therefore the pursuit command could not be realized on a velocity basis. A small leakage factor within the integrator was intended to use this anti-drift component for pursuit, e.g. for having the camera smoothly following changes in the line of sight. Thus the EUROCOPTER system corresponds to the AUDI system without the pursuit command in Fig. 6.6.

6.2.4 Test Flights

To prove the operability and evaluate the performance of the camera stabilization in helicopter applications, test flights were conducted at the EUROCOPTER airfield in Donauwörth/Germany in late 2006.

The test program consisted of two different types of maneuvers. The first step was aimed to evaluate the basic functions of camera stabilization, test the operability over the whole range of intended helicopter motion and assess the performance of the adaptive part of the algorithm. The second series then targeted the actual application, providing the pilot and the operator with an image of the landing place during the entire final approach.

Fig. 6.20 plots data from a test where the pilot induced pitch motion, starting with an amplitude of $\pm 10°$ and constantly increasing up to $\pm 30°$. The *top left* graph shows the time history of the camera tilt angle, which corresponds to pitch motion. Since a point of interest on the Earth was targeted, the camera has a bias of about $10°$ downwards. With little exceptions, mainly at about $t = 12s$ and $t = 46s$, no major saccades were necessary to keep the POI on the camera image. The fact that these saccades are very short-term obstructions of the otherwise sinusoidal motion indicates that the inaccuracy is a result of variances in the navigation system. If the GPS determines the position of the helicopter with some noisy variation, this leads to a sudden jump in the line of sight, resulting in a saccade. Since the point of interest was not straight in front of the helicopter, pitch motion required a compensation in the roll axis to keep the camera image free of rotation (Fig. 6.20 *top right*).

Especially during the second half of the test, the pilot increased the maximum pitch rate and the angular acceleration when stopping the motion. This required significant changes in the orientation of the helicopter rotor, inducing strong vibrations clearly visible in the gyroscope output at $t = 25s$, $t = 34s$ and $t = 43s$. The same applies to accelerations, exemplarily displayed for the yaw axis in the bottom right graph. What is remarkable is the generally high level of vibration, which will be discussed below.

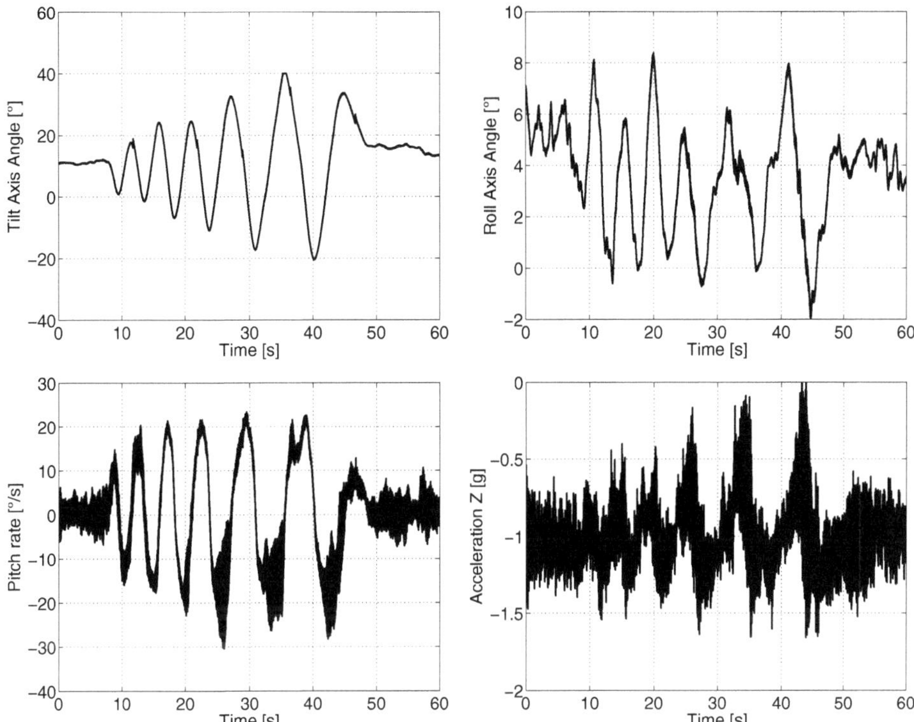

Fig. 6.20: Camera tilt and roll angle during increasing pitch motion (*top* row); corresponding measured pitch rate and vertical acceleration (*bottom* row).

As mentioned above, the FORBIAS system was connected to the Pilas software for providing an image of the landing place while guided by a Pilas flight tunnel in the final approach. Fig. 6.21 shows the evolution of the camera angles. Since there are obvious variances, the desired tilt (pitch) angle, as calculated by the helicopter navigation, and the actual camera tilt angle are plotted in the left column.

After a short pitching up of the helicopter during the first $10s$ of the test, the pilot went on a smooth flight path. When approaching the airfield and flying over it without actually touching ground, the camera tilt angle significantly increases to stay focused on the programmed landing point, especially in the second part of the test.

At two time intervals an increased number of saccades are remarkable. Between $t = 13s$ and $t = 15s$ the gyroscope corresponding to the tilt axis shows a bias shift of $4°/s$, leading to drift that is then corrected with saccades. During this interval all accelerometers measured doubled high-frequency vibration. Since only the tilt axis drifted, this indicates that the internal vibration compensation of that individual gyroscope was not properly trimmed by the manufacturer. The second set of saccades occur when approaching the point of interest during the second half of the test cycle. Here the relative approach velocity, e.g. the ratio of helicopter velocity and distance to the POI, is higher. During that period the position-based control without velocity-based pursuit cannot cope with the required pursuit demands, compensation is achieved with saccades. For the general impression on the pilot and the test engineer, these saccades had only little influence.

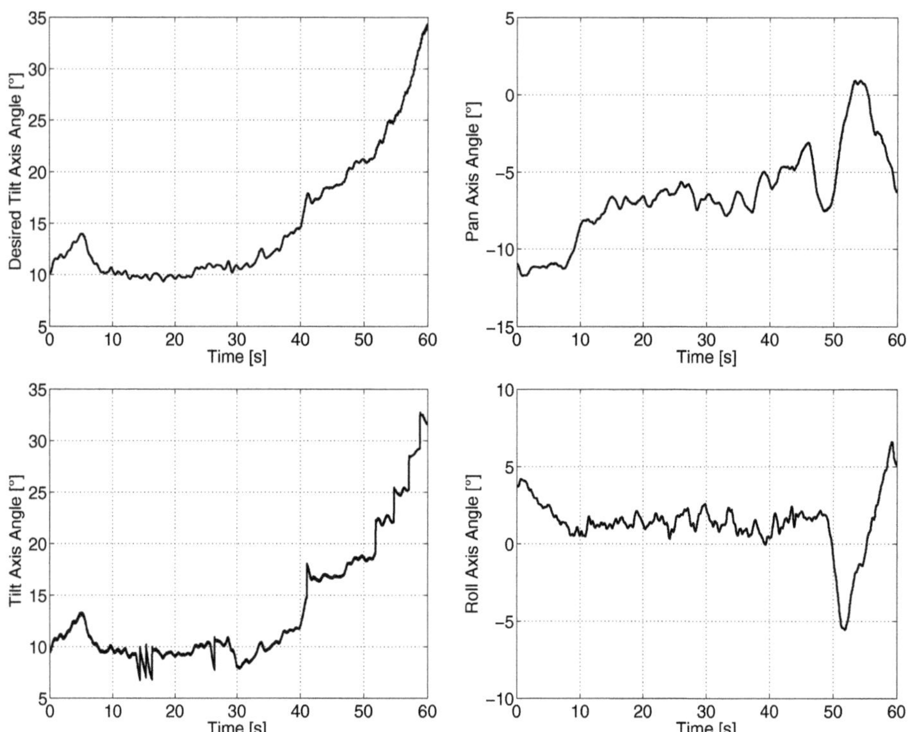

Fig. 6.21: Desired and actual tilt angle (*left*), actual pan and roll angle of the camera during a final approach (*right*).

During the last $15s$ the helicopter turned around the yaw axis. To compensate for this motion, a combined yaw-roll rotation of the camera is necessary due to the cardanic structure of the camera motion device.

The most critical part of camera stabilization in helicopters is the dynamic range, or the frequency spectrum of excitations. Fig. 6.22 exemplarily shows the Fast Fourier Transform of the pitch gyroscope data. In accordance with information from the

EUROCOPTER design office, the first eigenfrequency (EF) is distinctively measured at $26.5Hz$. Higher frequencies are negligible in comparison, but the effects of this first EF have to be reduced as discussed in Section 6.2.5.

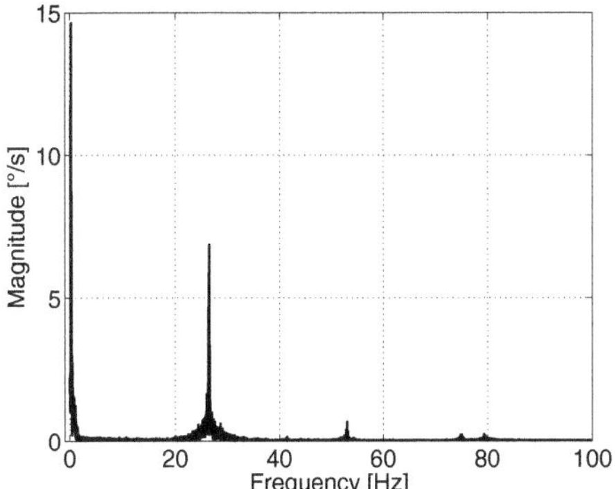

Fig. 6.22: Fast Fourier Transform of the pitch gyroscope, showing a distinct first eigenfrequency of $26.5Hz$.

To give a better impression of the test flight, Fig. 6.23 shows two photographs taken in-flight. The image of the FORBIAS camera is displayed on the central monitor between the pilot and the flight engineer. In addition, the Pilas flight path tunnel is displayed directly in front of the pilot to assist keeping the helicopter on track. The right photograph shows how the image of the landing place is presented to the pilot. In this case, the beginning of the helipad is programmed to be the target for the stabilized camera.

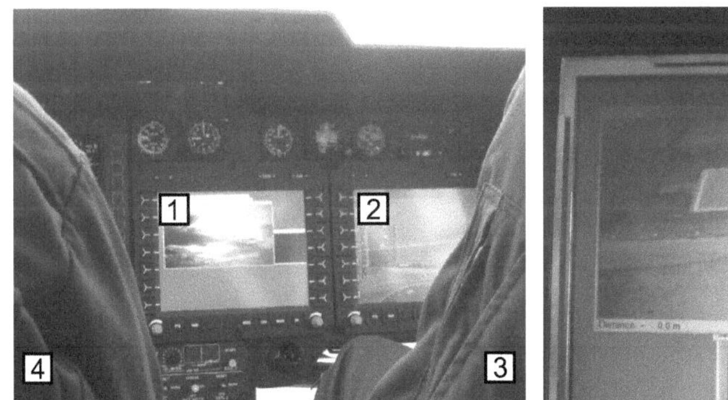

Fig. 6.23: In-flight photographs of the cockpit (*left*) with FORBIAS screen (1), Pilas screen (2), pilot (3) and flight engineer (4); the beginning of the helipad is the programmed POI shown to the pilot (*right*).

6.2.5 Discussion of Results

In the debriefing after the test flights, both the pilot and the flight engineer expressed their appreciation for the subjectively very stable camera image. Even flight maneuvers normally not occurring during regular flights could not degrade the quality of camera stabilization. During all final approaches the camera tracked the landing place to their high satisfaction. In their opinion, the saccades explained in connection with Fig. 6.21 did not significantly disturb the general impression.

As mentioned above, the comparatively high first eigenfrequency of the helicopter requires additional attention. With its $26.5Hz$ being far beyond the $6 - 8Hz$ cut-off frequency of the camera motion device, it cannot be compensated for with the current setup. There are two main possibilities to improve the system. A camera motion device specifically optimized for the most recent version of the camera could reach a significantly higher bandwidth. The second option is a more effective passive vibration isolation. However, this can only be reached by increasing the mass of the camera platform, which is usually not an option in aeronautics, or by decreasing the stiffness of the suspension. Due to stability reasons, the latter can only be achieved to a certain extent. Thus, a combination of both methods will be required for future applications in helicopters.

The influence of high frequency vibrations can be seen in the stabilization performance graphs in Fig. 6.24 for the tilt axis. Equivalent to the automotive tests, the stabilized camera is compared to the case when the CMD actuators are switched off. Due to the higher angular rates in comparison with the car motion, the effects become more obvious.

In the non-stabilized case (*left*), the pitch motion of the helicopter results in a significant optic flow in tilt direction, directly corresponding to the sinusoids of angular rates. With the FORBIAS system in full operation (*right*), the residual optic flow could be reduced by more than 80% in the early stage of the test. The effects of high frequency vibrations due to increased angular accelerations are remarkable, as mentioned in the previous section. These vibrations, visible in the second half of the test cycle, have a direct influence on the optic flow, increasing the flow level up to 8 pixel/frame. This emphasizes the need for measures to cope with the first eigenfrequency.

Notable also is the generally high level of vibrations. Without intended flying maneuver, the camera platform is subject to rotation at up to $8°/s$, in comparison to $2°/s$ in the automotive application. This shows the necessity of camera stabilization, especially in connection with high focal length lenses.

As a quality indicator for the performance of the camera stabilization, the correlation of optic flow and angular rates is evaluated, equivalent to the test drives.

In Fig. 6.25 (*left*) the correlation for the pitch axis is displayed. The regression line for the stabilized camera has close to no slope, indicating independency of optic flow and angular rates, but the general level of residual flow is still not negligible for the above mentioned reasons. In comparison to the fixed camera, the sign of the slope is reversed, meaning that there is a slight overcompensation in helicopter motion.

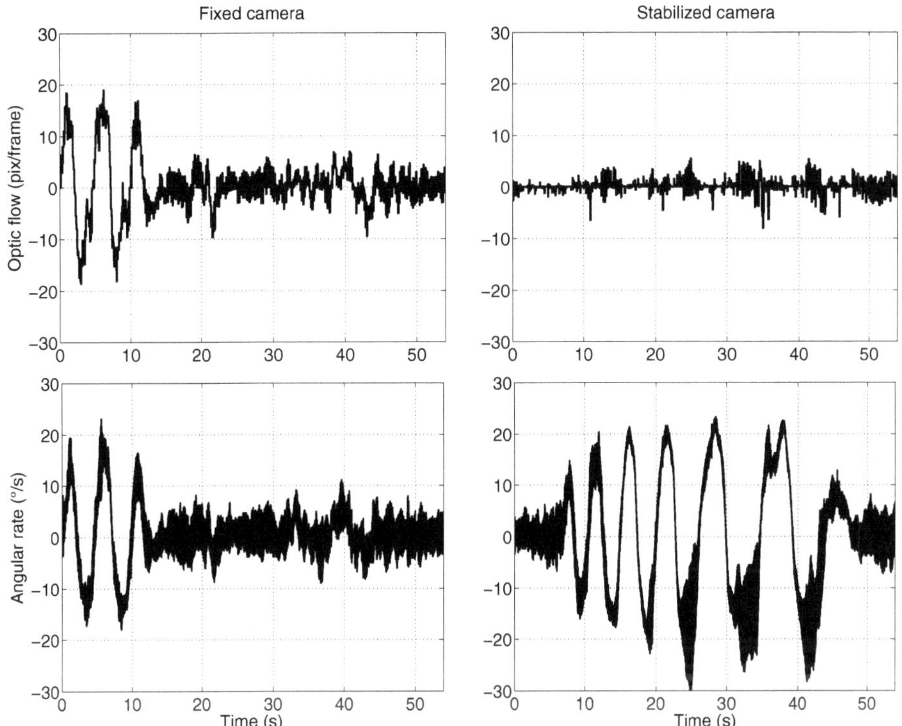

Fig. 6.24: Optic flow and angular rates during test flights with a helicopter-fixed (*left*) and an inertially stabilized camera (*right*).

The opposite is observable for the yaw axis (Fig. 6.25, *right*). In addition to a residual optic flow higher than for the pitch axis, the correlation could not be removed completely. Since the slope of the regression line has the same sign as in the unstabilized case, the system undercompensates for helicopter motion.

The fact that stabilization in pitch/tilt direction is better than in yaw/pan direction is remarkable in close to all tests, as well as in the initial tests on the Steward-Gough platform. Additionally the pan axis shows a wider spread of measurement points around the regression line, indicating more random optic flow. Since the helicopter camera pictured a quasi-static environment, this cannot be the result of moving objects as may be the case during test drives. Unlike the tilt axis, where the frame of the camera motion device is directly connected to the drive side of the gear box, the camera suspension is connected to the pan actuator via a push rod and two ball joints. Experience with the CMD showed that these joints loosen up after some time of operation, leading to backlash that results in low but random camera motion in pan direction.

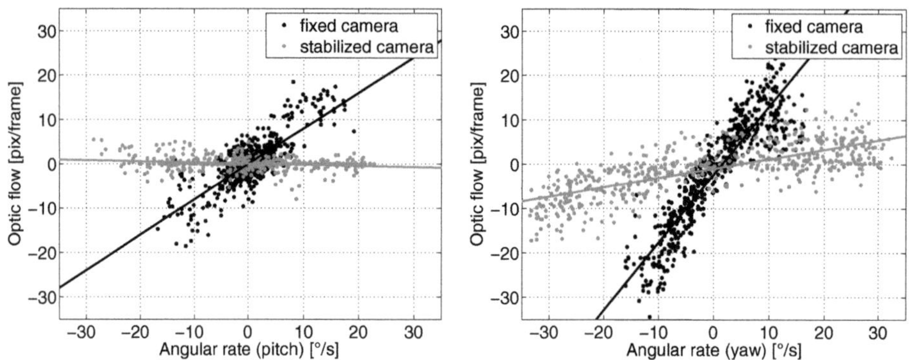

Fig. 6.25: Decorrelation of optic flow and helicopter motion in pitch (*left*) and yaw (*right*) direction.

Over- and undercompensating in the different axes is reflected in the adaptation of sensor gains. In general, the 60 second test cycles, or even the complete 15 minute test program, are not long enough for the complete system adaptation presented in the previous chapter. Therefore the system was pre-adapted under laboratory conditions. Due to different environmental conditions, the determined parameters could not be optimal for the actual test flight. Deviations from the optimum lead to the residual correlation shown in Fig. 6.25.

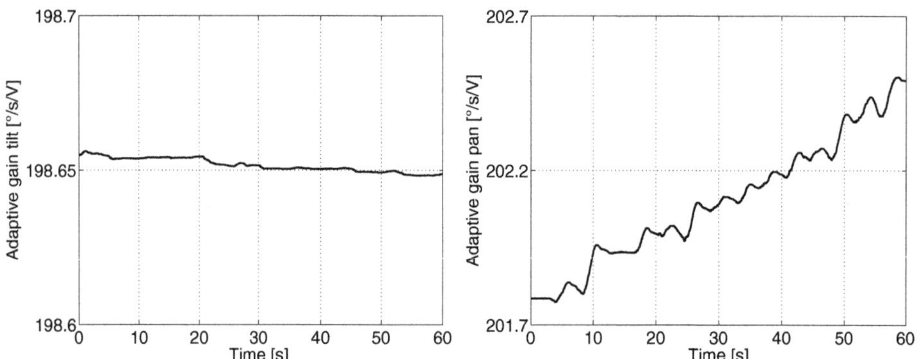

Fig. 6.26: Sensor gain pitch (*left*) and yaw (*right*).

During the test flights, at least the qualitative behavior of adaptation could be observed. In Fig. 6.26 the evolution of the sensor gains corresponding to pitch (*left*) and yaw (*right*) are plotted. With the pitch axis only slightly overcompensating, a slow decrease in the sensor gain can be observed. In yaw direction, however, the deviation from optimal values is more important and the motion compensation is not sufficient. Hence the sensor gain increases, and due to the inferior starting conditions at a rate about 80 times higher, showing that the in-flight adaptation works well and converges after sufficient time.

The duration of this adaptation can be influenced by the learning rate. Although more time is required, the choice of a low learning rate increases stability and is generally preferable.

6.2.6 Recommendations for Future Applications

After the complete design and evaluation of the camera stabilization system in successful test flights, further research towards possible future applications will facilitate strategical decisions. To quantify the amount of value added during night flights, test flights with an infrared sensitive camera should be performed. This would allow researchers to see whether the system is useful for the pilot during the final approach. In addition, the applicability for detecting people should be checked to verify whether the FORBIAS system can replace high-cost stabilized cameras currently used by Federal Border Patrols.

In the tests presented in this chapter, the stabilized camera was equipped with lenses with focal lengths (FL) of $16mm$ and $25mm$. To evaluate the performance with larger distances to the point of interest, it will be useful to test lenses with an FL of up to $50mm$.

In addition to these strategic tests, some technical alterations seem promising. In order to avoid frequent saccades during the last part of the final approach, a velocity-based pursuit command, as in the automotive application, should be implemented. The algorithm presented in Section 6.1.4 can be directly used without changes when the helicopter velocity with respect to the Earth is provided by the helicopter navigation system.

With knowledge of the frequency spectrum of vibrations, the combination of a wider bandwidth of the camera motion device and additional passive vibration isolation should further reduce the influence of the first eigenfrequency.

And finally, as a point brought up by the test pilot, the image of the stabilized camera in the cockpit should be full-screen on the monitor for easy recognition of obstacles.

7 Summary and Conclusion

Cognitive assistance systems are an important part of current research. On the academic side, its importance has recently been underlined with the funding of the Cluster of Excellence *Cognition for Technical Systems* by the *German Research Foundation*. Industrial relevance is obvious from the degree of effort companies, especially in the automotive field, make to become technological leaders [87]. Beyond that, many aeronautical applications are emerging.

Driver assistance systems tend towards cognition for improving comfort and safety. Only if the vehicle is completely aware of its own state as well as its vicinity, including traffic, can intelligent decisions for collision avoidance or mitigation be taken. In a future application, completely autonomous driving can be considered.

This work focuses on the use of inertial measurement units for cognitive systems. The focus lies on two main applications, orientation determination and camera stabilization.

The first topic was addressed with multi-sensor fusion algorithms. In contrast to aeronautical navigation applications, low-cost MEMS sensors required more sophisticated error compensation methods. In addition to the complementary filter, which is restricted to pitch/roll angle output for implementation on embedded microcontrollers, three stochastic fusion algorithms have been presented. The commonly used framework of Extended Kalman Filtering was compared to two theoretical approaches that are in the focus of current research: Sigma-Point Kalman Filtering, and Particle Filters.

The Sigma-Point approach better accounts for non-linearities and shows a slightly better accuracy performance than the EKF. In experiments on the Steward-Gough platform of the Institute of Applied Mechanics, an rms error of about $1°$ in pitch and roll direction could be achieved. In yaw direction the high magnetic fields of the parallel robot led to an error of $2\text{-}7°$. Particle Filters require high computational power and thus are not yet suitable for real-time applications. Therefore SPKF proved to be the most suitable algorithm at the moment.

The second main application was inertial camera stabilization. A pivotable camera with telephoto lens was actively stabilized to compensate for the vehicle's self-motion. Furthermore, it was programmed to track points of interest. Gaze stabilization requires a high degree of precision in motion sensing and camera actuation. With sensor characteristics being guaranteed by the manufacturer only within large

W. Günthner: *Enhancing Cognitive Assistance Systems with Inertial Measurement Units*, Studies in Computational Intelligence (SCI) **105**, 131–132 (2008)
www.springerlink.com

tolerances, a biomimetic control algorithm was designed for the adaptive compensation of those uncertainties.

Knowledge from neuroscience helped identify the required information at each point of the model. Thus the concept is based on the vestibulo-ocular reflex which stabilizes the eyes in head of vertebrates. By using this adaptive algorithm, cost-intensive calibration and identification procedures can be avoided.

Industrial application of especially gaze stabilization involves specific requirements for the inertial measurement unit. In addition to cost-efficiency, size is a dominating restriction. Therefore a custom-designed IMU was developed. With a design envelope of only $15mm \times 15mm \times 16mm$, the patented μCube saves up to 85% space in comparison with commercially available units [21]. The dynamic behavior of the sensor units, e.g. the bandwidth, exceeds most other comparable IMUs. For angular rate sensing up to 100Hz, and for linear acceleration up to 640Hz can be realized. These characteristics open up a wide market for inertial sensing beyond application in vehicles.

Fig. 7.1: Final design of the 6 dof inertial measurement unit μCube.

Finally the complete camera stabilization system, including inertial sensor, adaptive control and the camera motion device, was tested in two relevant industrial applications. In cooperation with Audi, the system was built into an A8 for test drives on German freeways. During these tests, points of interest like lane markers and other vehicles were tracked. Gaze stabilization removed all correlation between vehicle motion and residual optic flow in the camera image, which had been used as a quality criterion, and thus provided a stable image for image processing.

In the second application the stabilized camera was built into the nose of a helicopter. During the final approach, it delivered a stable image that helped the pilot and flight engineer assess a possible landing place from large distances. In a future application, this image can be used for automated obstacle detection.

Bibliography

[1] J.S. Albus. A theory of cerebellar function. *Mathematical Biosciences*, 10:25–61, 1971.

[2] C. Alt. Entwicklung eines miniaturisierten inertialmesssystems. Master's thesis, Lehrstuhl für Angewandte Mechanik, TU München, 2006.

[3] Analog Devices Inc. Adxl103/203 datasheet. www.analog.com, 2003.

[4] Analog Devices Inc. Adxrs150 datasheet. www.analog.com, 2003.

[5] Analog Devices Inc. Adxrs300 datasheet. www.analog.com, 2003.

[6] W.T. Ang, P.K. Khosla, and C.N. Riviere. Kalman filtering for real-time orientation tracking of handheld microsurgical instrument. In *Proc. of IEEE/RJS International Conference on Intelligent Robots and Systems*, Sendai, Japan, 2004.

[7] D. Angelaki and J.D. Dickmann. Gravity or translation: Central processing of vestibular signals to detect motion or tilt. *Vestibular Research*, 13:245–253, 2003.

[8] Atmel Corporation. Enhancing adc resolution by oversampling, avr121. www.atmel.com, September 2005.

[9] Atmel Corporation. At90can128 avr microcontroller datasheet. www.atmel.com, January 2007.

[10] A.-J. Baerveldt and R. Klang. A low-cost and low-weight attitude estimation system for an autonomous helicopter. In *Proc. of the IEEE International Conference on Intelligent Engineering Systems*, 1997.

[11] V. Blervaque, K. Mezger, L. Beuk, and J. Loewenau. Adas horizon: How digital maps can contriute to road safety. In J. Valdorf and W. Gessner, editors, *Advanced Microsystems for Automotive Applications*, pages 427–436. Springer, 2006.

[12] Bosch Sensortec GmbH. Angular rate sensor smg061 analog output datasheet. www.bosch-sensortec.com, 2006.

[13] Bosch Sensortec GmbH. Triaxial analog acceleration sensor smb363 datasheet. www.bosch-sensortec.com, 2006.

[14] D.M. Broussard and C.D. Kassardjian. Learning in a simple motor system. *Learn. Mem.*, 11:127–136, 2004.

[15] R.G. Brown and P.Y. Hwang. *Introduction to Random Signals and Applied Kalman Filtering*. John Wiley & Sons, 2001.

[16] U. Buettner. Das optokinetische system. In A. Huber and D. Koempf, editors, *Klinische Neuroophthalmologie*, pages 75–77. Thieme, 1998.

[17] Canon Kabushiki Kaisaha. Image stabilizing system. US Patent 6618197, September 2003.

[18] Canon Kabushiki Kaisaha. Image stabilization control device for use in camera system optionally including optical cahraceristics modifying converter. US Patent 6694096, February 2004.

[19] Canon Kabushiki Kaisaha. Image stabilizing optical lens device with decentering of lens sub-unit. US Patent 6728033, April 2004.

[20] M.J. Caruso. Applications of magnetoresistive sensors in navigation systems. Technical report, Honeywell Inc.

[21] Cloud Cap Technology, Inc. Crista inertial measurement unit (imu) interface/ operation document, May 2004.

[22] R.K. Curey, M.E. Ash, L.O. Thielman, and C.H. Barker. Proposed ieee inertial systems terminology standard an other inertial sensor standards. In *IEEE Postion Locaton Navigation Symposium*, Monterey, CA, April 2004.

[23] Cygnal Integrated Products, Inc. Improving adc resolution by oversampling and averaging. www.cygnal.com, May 2001.

[24] M. Dai, A. Klein, B. Cohen, and T. Raphan. Model-based study of the human cupular time constant. *J Vestib Res.*, 9:293–301, 1999.

[25] P. Dean, J. Porrill, and J. Stone. Visual awareness and the cerebellum: possible role of decorrelation control. *Progress in Brain Research*, 144:61–75, 2004.

[26] P. Dean, J. Porrill, and V. Stone. Decorrelation control by the cerebellum achieves oculomotor plant compensation in simulated vestibulo-ocular reflex. *Proc. R. Soc. Lond. B*, 269:1895–1904, 2002.

[27] N. Di Giusto. Evolution of new functionalities in vehicles enabled by the integration of microsystems: a route for european competitiveness. Keynote lecture at AMAA 2006, April 2006.

[28] E.D. Dickmanns. 4d-szenenanalyse mit integralen raumzeitlichen modellen. DAGM87, Informatik Fachberichte 149, 1987.

[29] E.D. Dickmanns. The 4d-approach to dynamic machine vision. In *Proc. of 33rd IEEE Conference on Decision and Control*, 1994.

[30] E.D. Dickmanns. An advanced vision system for ground vehicles. In *1st Workshop on In-Vehicle Cognitive Computer Vision Systems (IVC² VS) in conjunction with 3rd International Conference on Computer Vision Systems (ICVS)*, Graz, Austria, April 2003.

[31] G. Dissanayake, S. Sukkarieh, E. Nebot, and H. Durrant-Whyte. The aiding of a low-cost strapdown inertial measurement unit using vehicle model constraints for land vehicle applications. In *IEEE Transactions on Robotics and Automation*, volume 17, October 2001.

[32] R. Dorf and R. Bishop. *Modern Control Systems*. Prentice Hall, 2004.

[33] R. Dorobantu. Simulation des verhaltens einer low-cost strapdown imu unter laborbedingungen. Schriftenreihe des Instituts für Astronomische und Physikalische Geodäsie, TU München, 1999.

[34] A. Doucet, N. de Freitas, and N. Gordon. *Sequential Monte Carlo Methods in Practice*. Springer Science+Business Media, Inc., 2001.

[35] A.A.J. Egmond, J.J. Groen, and L.B.W. Jongkees. The mechanics of the semicircular canal. *Physiol.*, 110:1–17, 1949.

[36] S.J. Elliott and P.A. Nelson. Active noise control. *IEEE Signal Processing Magazine*, pages 12–35, 1993.

[37] C. Fernandez and J.M. Goldberg. Physiology of peripheral neurons innervating semicircular canals of squirrel money. i. *Neurophysiol.*, 34:635–660, 1971.

[38] C. Fernandez and J.M. Goldberg. Physiology of peripheral neurons innervating semicircular cans of squirrel monkey. ii. *Neurophysiol*, 34:661–675, 1971.

[39] FORBIAS. Internal minutes of meeting, July 2004.

[40] FORBIAS. Jahresbericht forschungsverbund bioanaloge sensomotorische assistenz, 2005.

[41] FORBIAS. Jahresbericht forschungsverbund bioanaloge sensomotorische assistenz, 2006.

[42] M. Frigo and S.G. Johnson. Fftw: A adaptive software architecture for the fft. In *Proc. of the International Conference on Acoustics, Speech, and Signal Procssing*, 1998.

[43] G.A. Geen. Progress in integrated gyroscopes. *IEEE Aerospace and Electronic Systems Magazine*, 19(11):12–17, 2004.

[44] J.A. Geen. Very low cost gyroscopes. In *Proc. of IEEE Sensors 2005*, 2005.

[45] J.A. Geen, S.J. Sherman, J.F. Chang, and S.R. Lewis. Single-chip surface micromachined integrated gyroscope with 50 deg/s allan deviation. *IEEE Journal of solid-state circuits*, 37:1860–1866, 2002.

[46] H. Geitner. Accelerometes decrease power consumption and increase reliability of washing machines. In *Proc. of IATC*, Chicago, USA, March 2006.

[47] M. Gienger. *Entwurf und Realisierung einer zweibeinigen Laufmaschine*, volume Nr. 378 of *Reihe 1*. VDI-Verlag, Düsseldorf, 2005.

[48] M. Gienger, K. Löffler, and F. Pfeiffer. Walking control of a biped robot based on inertial measurement. In *Proceedings of the International Workshop on Humanoid and Humanfriendly Robotics (IARP)*, Tsukuba, Japan, 2002.

[49] S. Glasauer. Cerebellar contribution to saccades and gaze holding: a modeling approach. *Ann NY Acad Sci*, 1004:206–219, 2003.

[50] S. Glasauer. Mathematische modellierung und analyse vestibulärer und okulomotorischer funktion bei normalpersonen und patienten. Habilitationsschrift, LMU München, 2004.

[51] S. Glasauer. Vestibular and motor processing fo head direction signals. In SI Wiener and JS Taube, editors, *Head direction cells and the neural mechanisms of spatial orientation*. MIT Press, 2005.

[52] S. Glasauer, M. Dieterich, and T. Brandt. Central positional nystagmus simulated by a mathematical ocular motor model of otolith-dependent modification of listings plane. *Neurophysiol*, 86:1546–11554, 2001.

[53] S. Glasauer, M. Hoshi, and U. Büttner. Smooth pursuit in patients with downbeat nystagmus. *Ann NY Acad Sci*, 1039:532–535, 2005.

[54] S. Glasauer and D.M. Merfeld. Modelling three dimensional vestibular responses during complex motion stimulation. In M. Fetter, T. Haslwanter, and H. Misslisch, editors, *Three-dimensional kinematics of eye, head and limb movements*, pages 387–398. Harwood academic publishers, 1997.

[55] W. Günthner, S. Glasauer, P. Wagner, S. Bardins, and H. Ulbrich. Technical realisation of adaptive sensor and plant compensation for three-dimensional gaze stabilisation. In *Society for Neuroscience, No. 391.8*, Washington DC, USA, November 2005.

[56] W. Günthner, S. Glasauer, P. Wagner, S. Bardins, and H. Ulbrich. A highly miniaturized inertial measurement unit for biomimetic gaze stabilization. In *Proc. of The Second International Conference on Dynamics*, Beijing, China, 2006.

[57] W. Günthner, S. Glasauer, P. Wagner, and H. Ulbrich. Biologically inspired multi-sensor fusion for adaptive camera stabilizationi n driver-assistance systems. In J. Valdorf and W. Gessner, editors, *Advanced Microsystems for Automotive Applications*, pages 107–121. Springer, 2006.

[58] W. Günthner, S. Glasauer, P. Wagner, and H. Ulbrich. Biomimetic control for adaptive camera stabilization in driver-assistance systems. In *Proc. of The 8th International Conference on Motion and Vibration Control MOVIC*, Daejeon, Korea, 2006.

[59] W. Günthner and H. Ulbrich. Miniaturisiertes intertialmesssystem. Offenlegungsschrift DE 10 2005 047 873, Deutsches Patent- und Markenamt, April 2007.

[60] W. Günthner, Ph. Wagner, and H. Ulbrich. An inertially stabilized vehicle camera system - hardware, algorithms, test drives. In *Proc. of IEEE IECON'06, IEEE Industrial Electronics Society*, Paris, France, 2006.

[61] N.J. Gordon. Non-linear/non-gaussian filtering and the bootstrap filter. In *Proc. of IEE Colloquium on Non-Linear Filters*, 1994.

[62] H. Gray. *Anatomy of the Human Body*. Lea & Febiger, Philadelphia, 1918.

[63] L. Hakansson. The filtered-x lms algorithm. Adaptive Signal Processing Classnotes ETC004, January 2006.

[64] S. Haykin. *Adaptive Filter Theory*. Prentice-Hall, 2002.

[65] S. Highstein, J. Porrill, and P. Dean. Report on a workshop concerning the cerebellum and motor learning held in st. louis. In *The Cerebellum*, 2004.

[66] J.E. Holly. Baselines for three-dimensional perception of combined linear and angular self-motion with changing rotational axis. *Vestibular Research*, 10:163–178, 2000.

[67] J.E. Holly and G. McCollum. The shape of self-motion perception - i. equivalence classification for sustained motions. *Neuroscience*, 70(2):461–486, 1996.

[68] J.E. Holly and G. McCollum. The shape of self-motion perception - ii. framework and principles for simple and complex motion. *Neuroscience*, 70(2):487–513, 1996.

[69] Honeyell. 1, 2 and 3 axis magnetic sensors hmc1051, hmc1052, hmc1053. www.honeywell.com.

[70] Honeywell SSEC. Set/reset function for magnetic sensors, applicaton note an213. www.ssec.honeywell.com.

[71] M. Huterer and K. Cullen. Vestibuloocular reflex dynamics during high-frequency and high-acceleration rotations of the had on body in rhesus monkey. *Neurophysiol*, 88:13–28, 2002.

[72] IGI Ingenieur-Gesellschaft für Interfaces mbH. Aerocontrol-iid performance test report, May 2005.

[73] IGI Ingenieur-Gesellschaft für Interfaces mbH. AEROcontrol Precise Positioning & Attituted Determination, June 2006.

[74] iMAR Gesellschaft für inertiale Mess-, Automatisierungs- und Regelsystem mbH. Vertical reference unit with fiber optic gyros, servo accelerometers and integrated strapdown processor, 2005.

[75] INNOVATE! Mission possible. www.innovate-magazn.de, March 2006.

[76] M. Ito. Cerebellar control of the vestibulo-ocular reflex - around the flocculus hypothesis. *Ann. Rev. Neurosci.*, 5:275–296, 1982.

[77] M. Ito. Cerebellar learning in the vestibulo-ocular reflex. *Trends in Cognitive Sciences*, 2(9):313–321, 1998.

[78] S.J. Julier and J.K. Uhlmann. A new extension of the kalman filter to nonlinear systems. In *Proc. of SPIE AeroSense Symposium*, Orlando, FL, April 1997.

[79] S.J. Julier and J.K. Uhlmann. Unscented filtering and nonlinear estimation. *Proceedings of the IEEE*, 92:1958–1958, 2004.

[80] JUMA Leiterplattentechnologie Markus Wölfel GmbH. Drahtgeschriebene leiterplatte oder platine mit leiterdrähten mit rechteckigem oder quadratischem querschnitt. Gebrauchsmusterschrift DE 20 2005 001 161 U1, May 2005.

[81] JUMA PCB GmbH. Verfahren zur herstellung einer räumlichen leiterplattenstruktur aus wenigstens zwei zueinander winkelig angeordneten leiterplattenabschnitten. Offenlegungsschrift DE 10 2005 003 369 A1, July 2006.

[82] R.E. Kalman. A new approach to linear filtering and prediction problems. *Basic Engineering*, 82:35–45, 1960.

[83] E.R. Kandel, J.H. Schwartz, and T.M. Jessell. *Principles of Neural Science*. McGraw-Hill Mdeical, 2000.

[84] C. T. Kelley. Unconstrained implicit filtering code: Implicit filtering with sr1 and bfgs quasi-newton methods. Source code online, http://www4.ncsu.edu/ ctk/darts/imfil.m, 1998.

[85] C.T. Kelley. Users guide for imfil version 0.5. http://www4.ncsu.edu/ ctk/, November 2005.

[86] R.E. Kettner, S. Mahamud, H.-C. Leung, N. Sitkoff, J.C. Houk, B.W. Peterson, and A.G. Barto. Prediction of complex two-dimensional trajectories by a cerebellar model of smooth pursuit eye movement. *Neurophysiol.*, 77(4):2115–2130, 1997.

[87] A. Kirchner. Driver assistance - steps towards the seeing car. Advanced Microsystems for Automotive Applications, 2007.

[88] J.B. Kuipers. *Quaternions and rotation sequences*. Princeton Univ. Press, 2002.

[89] S. Kuo and D. Morgan. *Active Noise Control*. John Wiley and Sons, Inc., 1996.

[90] Eric A. Lehmann and Robert C. Williamson. Particle filter design using importance sampling for acoustic source localisation and tracking in reverberant environments. *EURASIP Journal on Applied Signal Processing*, 2006.

[91] R. J. Leigh and D. S. Zee. *The Neurology of Eye Movements*. Oxford University Press, New York, Oxford, 2006.

[92] S.G. Lisberger. Physiologic basis for motor learning in the vestibulo-ocular reflex. *Otolaryngol Head Neck Surg*, 119:43–48, 1998.

[93] J. Loevenau, K. Gresser, D. Wisselmann, W. Richter, and D. Rabel. Dynamic pass prediction: A new driver asistance system for superior and safe overtaking. In J. Valdorf and W. Gessner, editors, *Advanced Microsystems for Automotive Applications*, pages 67–78. Springer, 2006.

[94] D. Marr. A theory of cerebellar cortex. *Journal of Physiology*, 202:437–470, 1969.

[95] S. Marti, D. Straumann, U. Büttner, and S. Glasauer. A model-based theory on the origin of downbeat nystagmus. 2007, submitted.

[96] S. Marti, D. Straumann, and S. Glasauer. The origin of downbeat nystagmus: an asymmetry in the distribution of on-directions of vertical gaze-velocity purkinje cells. *Ann NY Acad Sci*, 1039:548–553, 2005.

[97] Maxim Integrated Products. Max1226 datasheet. www.maxim-ic.com, July 2004.

[98] P. Maybeck. *Stochastic Models, Estimation, and Control*, volume 1. Academic Press, 1979.

[99] T. Mergner and S. Glasauer. A simple model of vestibular canal-otolith signal fusion. *Annals New York Academy of Sciences*, 871:430–434, 1999.

[100] J. Metzger, K. Wisotzky, J. Wendel, and G.F. Trommer. Sigma-point filter for terrain referenced navigation. In *Proc. of the AIAA Guidance, Navigation, and Control Conference*, 2005.

[101] F.A. Miles and S.G. Lisberger. Plasticity in the vestibulo-ocular reflex: a new hypothesis. *Ann. Rev. Neurosci.*, 4:273–299, 1981.

[102] Murata Manufacturing Co., Ltd. Piezoelectric vibrationg gyroscopes datasheet. www.murata.com, 2003.

[103] S. Neumaier and G. Färber. Videobasierte fahrspurerkennung zur umfelderfassung bei straßenfahrzeugen. In *Proc. of AMS - Autonome Mobile Systeme*, 2005.

[104] S. Neumaier, Ph. Harms, and G. Färber. Videobasierte umfelderfassung zur fahrerassistenz. In *Proc. of 4. Workshop Fahrerassistenzsysteme FAS*, 2006.

[105] C.M. Oman, E.N. Marcus, and I.S. Curthoys. The influence of semicircular canal morphology on endolymph flow dynamics. *Acta Otolaryngol*, 103:1–13, 1987.

[106] F. Panerai, G. Metta, and G. Sandini. Adaptive image stabilization: a need for vision-based active robotic agents. In *Proc. of SAB 2000*, Paris, France, September 2000.

[107] F. Panerai, G. Metta, and G. Sandini. Learning vor-like stabilization reflexes in robots. In *Proc. of the 8th European Symposium on Artificial Neural Networks*, 2000.

[108] F. Panerai, G. Metta, and G. Sandini. Learning visual stabilization reflexes in robots with moving eyes. *Neurocomputing*, 48:323–337, 2002.

[109] F. Panerai and G. Sandini. Oculo-motor stabilization reflexes: integration of inertial and visual information. *Neural Networks*, 11:1191–1204, 1998.

[110] A.S. Paul and E.A. Wan. Dual kalman filters for autonomous terrain aided navigation in unknown environments. In *Proceedings IEEE International Joint Conference on Neural Networks*, 2005.

[111] J. Porrill, P. Dean, and J. Stone. Recurrent cerebellar architecture solves the motor-error problem. *Proc. R. Soc. London B*, 271:789–796, 2004.

[112] J. Raymond and S. Lisberger. Neural learning rules for the vestibulo-ocular reflex. *Neuroscience*, 18(21):9112–9129, 1998.

[113] G. Reymond, J. Droulez, and A. Kemeny. Visuovestibular perception of self-motion modeled as a dynamic optimization process. *Biological Cybernetics*, 87:301–314, 2002.

[114] S. Riebe. Hexapod-diss, titel nachschlagen. Dissertation am Lehrstuhl für Angewandte Mechanik, TUM, 2006.

[115] S. Riebe and H. Ulbrich. Modelling and online-computation of the dynamics of a parallel kinematic with six-degrees-of-freedom. *Archive of Applied Mechanics*, 72:817–829, July 2003.

[116] B. Ristic, S. Arulampalam, and N. Gordon. *Beyond the Kalman Filter - Particle Filters for Tracking Applications*. Arech House Publishers Boston/London, 2004.

[117] R. Rodloff. Gibt es den optischen superkreisel? *Flugwiss. Weltraumforschung*, 18:2–15, 1994.

[118] A. Schmidt. Bestimmung des dynamischen verhaltens eines miniaturisierten inertialmesssystems. Semesterarbeit am Lehrstuhl für Angewandte Mechanik, January 2007.

[119] E. Seifert and A. Nauda. Enhancing the dynamic range of analog-to-digital converters by reducing excess noise. In *Proc. of the IEEE Pacific Rim Converence on Communications, Computers and Signal Processing*, 1989.

[120] SensoNor AS. Sar10 datasheet. www.sensonor.com, September 2003.

[121] T. Shibata and S. Schaal. Biomimetic gaze stabilization based on feedback-error-learning with nonparametric regression networks. *Neural Networks*, 14(2):201–216, 2001.

[122] T. Shibata, H. Tabata, S. Schaal, and M. Kawato. A model of smooth pursuit in primates based on learning the target dynamics. *Neural Networks*, 18:213–224, 2005.

[123] E.-H. Shin and N. El-Sheimy. An unscented kalman filter for in-motion alignment of low-cost imus. In *Proc. of the Position and Navigation Symposium PLANS*, 2004.

[124] Silicon Designs, Inc. Model 1221 low noise analog accelerometer datasheet. www.silicondesigns.com, August 2006.

[125] ST Microelectronics. Lis3lv02dq datasheet. www.st.com, 2005.

[126] M. Strupp and T. Brandt. Der vestibulookuläre reflex. In A. Huber and D. Koempf, editors, *Klinische Neuroophthalmologie*, pages 78–85. Thieme, 1998.

[127] Systron Donner Inertial. Mmq50 imu datasheet. www.systron.com.

[128] R.J. Telban. Investigation of mathematical models of otolith organs for human centered motion cueing algorithms. American Institute of Aeronautics and Astronautics, August 2000.

[129] S. Thrun and J. Kosecka. Lecture notes computer vision cs223b. www.stanford.edu, 2007.

[130] S. Thrun, M. Montemerlo, H. Dahlkamp, D. Stavens, A. Aron, J. Diebel, P. Fong, J. Gale, M. Halpenny, G. Hoffmann, K. Lau, C. Oakley, M. Palatucci, V. Pratt, P. Stang, S. Strohband, C. Dupont, L.-E. Jendrossek, C. Koelen, C. Markey, C. Rummel, J. van Niekerk, E. Jensen, P. Alessandrini, G. Bradski, B. Davies, S. Ettinger, A. Kaehler, A. Nefian, and P. Mahoney. Winning the darpa grand challenge. *Journal of Field Robotics*, 2006. accepted for publication.

[131] D. Titterton and J. Weston. *Strapdown inertial navigation technology*. The Institution of Electrical Engineers, 2nd ed. edition, 2004.

[132] R. van der Merwe and E.A. Wan. Sigma-point kalman filters for integrated navigation. In *Proceedings of the 60th Annual Meeting of The Institute of Navigation (ION)*, 2004.

[133] P. van der Smagt. Cerebellar control of robot arms. *Connection Science*, 10:301–320, 1998.

[134] P. van der Smagt and G. Hirzinger. The cerebellum as computed torque model. In *Fourth International Conference on Knowledge-Based Intelligent Engineering Systems*, 2000.

[135] P. Wagner. Systeme zur hochdynamischen blickrichtungssteuerung. Dissertation am Lehrstuhl für Angewandte Mechanik, TU München, 2007.

[136] P. Wagner, W. Günthner, and H. Ulbrich. Gaze control devices for driver assistance systems. In *Proceedings of The 8th International Conference on Motion and Vibration Control MOVIC*, 2006.

[137] P. Wagner, W. Günthner, and H. Ulbrich. Gaze stabilization: Technical realization of biomimetic eye motions. In *Proceedings of The Second International Conference on Dynamics*, Beijing, China, 2006.

[138] P. Wagner, W. Günthner, and H. Ulbrich. An inertially stabilised vehicle camera system for driver assistance applications. In *Proc. of MECHATRONICS 2006 - 4th IFAC-Symposium on Mechatronic Systems IFAC*, 2006.

[139] M. Wang, Y. Yang, R.R. Hatch, and Y. Zhang. Adaptive filter for a miniature mems based attitude and heading reference system. In *Proc. of IEEE Position Location and Navigation Symposium*, 2004.

[140] J. Wendel, A. Maier, J. Metzger, and G.F. Trommer. Comparison of extended and sigma-point kalman filters for tightly coupled gps/ins integration. In *Proc. of AIAA Guidance, Navigation, and Control Conference*, 2005.

[141] J. Wendel, R. Moenikes, J. Metzger, A. Maier, and G.F. Trommer. A performance comparison of tightly coupled gps/ins navigation systems based on extended and sigma point kalman filters. In *Proc. of ION GNSS*, 2005.

[142] B. Widrow and S. Stearns. *Adaptive Signal Processing*. Prentice-Hall, 1985.

[143] M. Wölfel. Verfahren und vorrichtung zur herstellung von drahteschriebenen leiterplatten. Patentschrift DE 196 18917 C1, October 1997.

[144] Xsens Technologies B.V.. Moven inertial motion capture datasheet, January 2007.

[145] L.R. Young. Perception of the body in space: mechanisms. In D. Smith, editor, *Handbook of Physiology - the Nervous System III*. American Physiological Society, 1984.

[146] P. Zhang, J. Gu, E.E. Milios, and P. Huynh. Navigation with imu/gps/digital compass with unscented kalman filter. In *Proc. of IEEE International Conference on Mechatronics and Automation*, 2005.

[147] C. Zhao. Probabilistic approaches for multi-sensor fusion in inertail measurement units. Diplomarbeit am Lehrstuhl für Angewandte Mechanik, April 2006.

[148] L.H. Zupan, D.M. Merfeld, and C. Darlot. Using sensory weighting to model the influence of canal, otolith and visual cues on spation orientation and eye movements. *Biol. Cybern.*, 86:209–230, 2002.